KU-589-348

There Were Giants Upon the Earth

ALSO BY ZECHARIA SITCHIN

The Earth Chronicles

The 12th Planet

The Stairway to Heaven

The Wars of Gods and Men

The Lost Realms

When Time Began

The Cosmic Code

The End of Days

The Earth Chronicles Handbook

Companion Books

Genesis Revisited

Divine Encounters

The Lost Book of Enki

Autobiographical Books

The Earth Chronicles Expeditions

Journeys to the Mythical Past

Cover illustration: Victory stela of king Naram-Sin, ca. 2250 B.C., now in the Louvre Museum, Paris.

There Were Giants Upon the Earth

Gods, Demigods, and Human Ancestry: The Evidence of Alien DNA

ZECHARIA SITCHIN

Bear & Company
Rochester, Vermont • Toronto, Canada

Bear & Company
One Park Street
Rochester, Vermont 05767
www.BearandCompanyBooks.com

Bear & Company is a division of Inner Traditions International

Copyright © 2010 by Zecharia Sitchin

All rights reserved. No part of this book may be reproduced or utilized in any form or by any means, electronic or mechanical, including photocopying, recording, or by any information storage and retrieval system, without permission in writing from the publisher.

Library of Congress Cataloging-in-Publication Data

Sitchin, Zecharia.
There were giants upon the Earth : gods, demigods, and human ancestry : the evidence of alien DNA / Zecharia Sitchin.
p. cm. — (Earth chronicles)
Summary: "The crowning work of the best-selling Earth chronicles series"—Provided by publisher.
ISBN 978-1-59143-121-3 (hardcover)
1. Civilization, Ancient—Extraterrestrial influences. 2. Extraterrestrial beings. 3. DNA. 4. Heredity. 5. Gods—History. 6. Goddesses—History. I. Title.
CB156.S585 2010
930—dc22

2010012959

Printed and bound in the United States by Lake Book Manufacturing

10 9 8 7 6 5 4

Text design and layout by Priscilla Baker
This book was typeset in Garamond Premier Pro, with Impact and Gill Sans used as display typefaces

Royal Tombs of Ur illustrations courtesy of the University of Pennsylvania Museum of Archaeology & Anthropology, Philadelphia, and the Trustees of the British Museum, London

To send correspondence to the author of this book, mail a first-class letter to P.O. Box 577, New York, NY 10185 or visit his website at **www.Sitchin.com.**

Contents

INTRODUCTION

And It Came to Pass

And it came to pass,
When men began to multiply on the face of the Earth
and daughters were born unto them,
that the sons of God saw the daughters of men
that they were fair, and they took them wives
of all which they chose.

There were giants upon the Earth
in those days and also thereafter too,
When the sons of God
came in unto the daughters of men
and they bare children to them—
the same Mighty Men of old,
Men of Renown.

The reader, if familiar with the King James English version of the Bible, will recognize these verses in chapter 6 of Genesis as the preamble to the story of the Deluge, the Great Flood in which Noah, huddled in an ark, was saved to repopulate the Earth.

The reader, if familiar with my writings, will also recognize these verses as the reason why many decades ago, a schoolboy was prompted to ask his teacher why it is "giants" who are the subject of these verses, when the word in the original Hebrew text is *Nefilim*—which, stemming from the Hebrew verb *NaFoL,* means to fall down, to be downed, to come down—and in no way 'giants'.

The schoolboy was I. Instead of being congratulated on my linguistic acumen, I was harshly reprimanded. "Sitchin, sit down!" the teacher hissed with repressed anger; "you don't question the Bible!" I was deeply hurt that day, for I was not questioning the Bible—on the contrary, I was pointing out the need to understand it accurately. And that was what changed my life's direction to pursue the *Nefilim.* Who were they, and who were their "Mighty Men" descendants?

The search for answers started with linguistic questions. The Hebrew text does not speak of "Men" who began to multiply, but of *Ha'Adam*—"***The Adam,***" a generic term, a human species. It does not speak of the sons of "God," but uses the term *Bnei Ha-Elohim*—the *sons* (in the plural) of ***The Elohim,*** a plural term taken to mean "gods" but literally meaning "The Lofty Ones." The "Daughters of The Adam" were not "fair," but *Tovoth*—good, compatible . . . And unavoidably we find ourselves confronting issues of *origins.* How did Mankind happen to be on this planet, and whose genetic code do we carry?

In just three verses and a few words—forty-nine words in the original Hebrew of Genesis—the Bible describes the creation of Heaven and Earth, then records an actual prehistoric time of early Mankind and a series of amazing events, including a global Flood, the presence on Earth of gods and their sons, inter-species intermarriage, and demigod offspring . . .

And so, starting with one word (*Nefilim*), I told the tale of the *Anunnaki,* "Those who from Heaven to Earth came"—space travelers and interplanetary settlers who came from their troubled planet to Earth in need of gold, and ended up fashioning The Adam in their image. In doing so I brought them to life—recognizing them individually, unraveling their tangled relationships, describing their tasks, loves, ambitions, and wars—and identifying their inter-species offspring, the 'demigods'.

I have been asked at times where my interests would have taken me were the teacher to compliment rather than reprimand me. In truth,

I have asked myself a different question: What if indeed "there were giants upon the Earth, in those days *and thereafter too*"? The cultural, scientific, and religious implications are awesome; they lead to the next unavoidable questions: Why did the compilers of the Hebrew Bible, which is totally devoted to monotheism, include the bombshell verses in the prehistoric record—and what were their sources?

I believe that I have found the answer. Deciphering the enigma of the demigods (the famed Gilgamesh among them), I conclude in this book—my crowning oeuvre—that compelling *physical evidence for past alien presence on Earth has been buried in an ancient tomb.* It is a tale that has immense implications for our genetic origins—a key to unlocking the secrets of health, longevity, life, and death; it is a mystery whose unraveling will take the reader on a unique adventure and finally reveal what was held back from Adam in the Garden of Eden.

ZECHARIA SITCHIN

I

Alexander's Quest for Immortality

In the spring of 334 B.C., Alexander of Macedon led a massive Greek army across the Hellespont, a narrow straits of water separating Europe from Asia (now called the Dardanelles), and launched the first known armed invasion of Asia from Europe. His military forces, with some 15,000 elite foot soldiers and cavalry, represented an alliance of Greek states formed in response to repeated invasions of Greece by the Asiatic Persians: First, in 490 B.C. (when the invasion was repulsed at Marathon) and then in 480/479 B.C., when the Persians humiliated the Greeks by occupying and sacking Athens.

The two sides have been warring since then over Asia Minor, where Greek settlements (of which Troy had been the most storied) were proliferating, and clashed over the lucrative sea lanes in the eastern Mediterranean. While the Persians were organized in a mighty empire ruled by a succession of "King of Kings," the Greeks were fragmented into quarreling city-states; the devastating and humiliating Persian invasions coupled with the continuing clashes on land and sea finally served as an impetus to form a League under the leadership of Macedonia; and the task of leading the counterattack was entrusted to Alexander.

He chose to cross from Europe to Asia at Hellespont ('A' on map, Fig. 1), the same narrows that the Persians had crossed for their invasions westward. In times past the narrows were dominated on the

Asian side by the fortified city of ***Troy***—the epicenter of the Trojan War that had raged there, according to Homer's *Iliad,* many centuries before. Carrying a copy of the epic tale given him by his tutor Aristotle, Alexander made it a point to stop at the ruins of Troy to offer sacrifices to the goddess Athena and pay homage at the tomb of Achilles (whose courage and heroism Alexander admired).

The crossing by the army of thousands was uneventful. The Persians, rather than ward off the invaders at the beach, saw a chance to annihilate the Greek force by luring it inland. A Persian army, led by one of their best generals, was waiting for Alexander and his army along

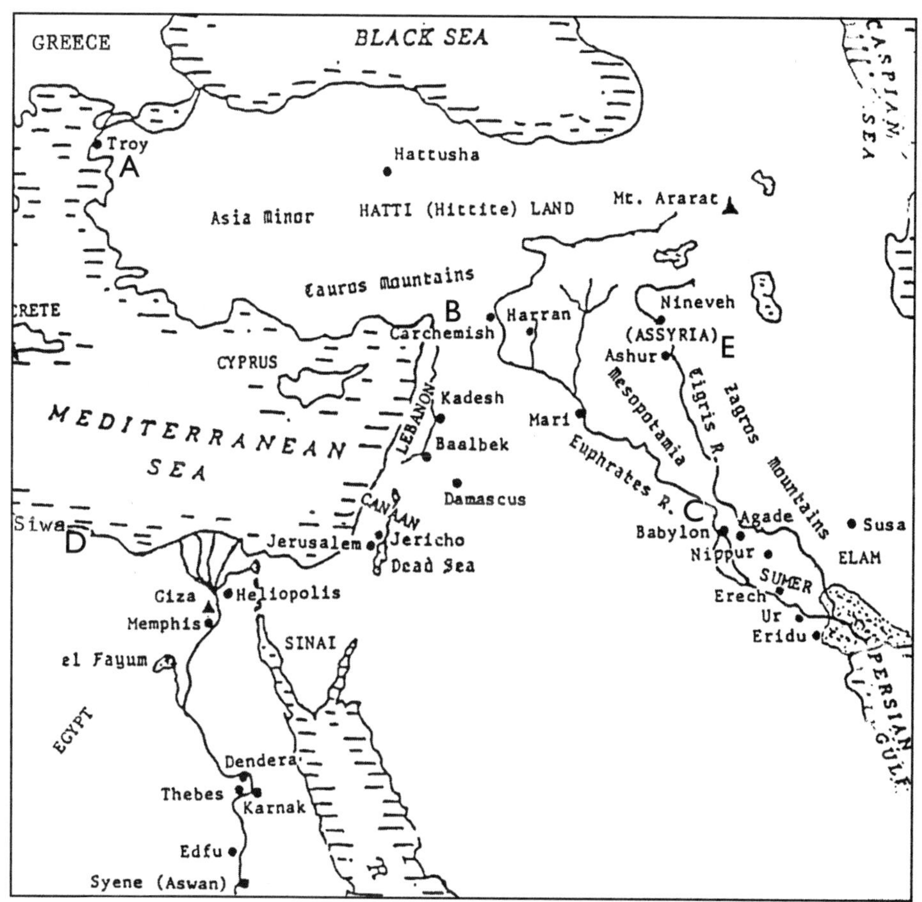

Figure 1

a river, forming a battle line somewhat inland; but though the Persians had the advantage of positions and numbers, the Greeks broke through. Falling back, the Persians assembled another army and even planned a counterinvasion of Greece; but in the meantime, their retreat enabled the Greeks to advance freely in Asia Minor, all the way to what is now the Turkish-Syrian border ('B' on map, Fig. 1).

In the fall of 333 B.C., the Persian *Shah-in-Shah* ("King of Kings") himself, Darius III, led a cavalry charge against Alexander's advancing troops; the battle, known as the battle of Issus (and much depicted by Greek artists, Fig. 2), ended with the capture of Darius's royal tent, but not of Darius himself. The Persian king, beaten but not defeated, retreated to Babylon ('C' on map, Fig. 1) the western headquarters of an empire that stretched from Asia Minor (where Alexander had invaded) all the way to India.

Incomprehensively, Alexander gave up the opportunity to crush the Persian enemy once and for all. Instead of pursuing the Persian remnants and their humbled king, he let Darius retreat eastward to Babylon and rouse the empire to continue the war. Giving up the chance for a decisive victory, Alexander instead set his course southward . . . The

Figure 2

defeat of the Persians to avenge their previous attacks on Greece—the reason for the Greek states' alliance under Alexander—was deferred to a later time. It was Egypt, not Persia, the astounded Greek generals discovered, that was the real pressing destination for Alexander.

What was on Alexander's mind, it was later revealed, was his own rather than Greece's destiny, for he was driven by persistent rumors in the Macedonian court that his real father was not King Philip, but a mysterious Egyptian. As related in various accounts, the court of King Philip was once visited by an Egyptian Pharaoh whom the Greeks called Nectanebus. He was a master magician, a diviner, and he secretly seduced Queen Olympias, Philip's wife; thus, though it was assumed when she gave birth to Alexander that it was King Philip who was the father, Alexander's true father was an Egyptian visitor.

These persistent rumors, which soured the relations between King Philip and the queen, gained credibility when Philip—some said in order to clear his way to marry a young daughter of a Macedonian nobleman—publicly accused Olympias of adultery, a step that cast doubt on Alexander's status as Crown Prince. It was perhaps then, but certainly not later than when the king's new wife was with child, that the story took another twist: The mysterious visitor who had presumably fathered Alexander was not a mere Egyptian—he was a god in disguise: the Egyptian god Amon (also spelled Ammon, Amun, Amen). According to this version, Alexander was more than a royal prince (the queen's son)—he was a demigod.

The issue of royal succession in Macedonia was settled when King Philip, reveling in the birth of a son by his new wife, was assassinated, and Alexander, at age 20, acceded to the throne. But the issue of his true parentage continued to engage Alexander; for if true, he was entitled to something more important than inheriting a royal throne—he was entitled to inherit the immortality of the gods!

With his accession to Macedonia's throne, Alexander replaced Philip as commander of the alliance of Greek states in their invasion project. But before embarking on the march to Asia, he made his way to Delphi, a distant sacred site all the way in southern Greece. It was the location of ancient Greece's most famed oracle to which kings and heroes went to consult about their future. There, in the temple to the god Apollo, a legendary priestess, the Sibyl, would go into a trance and, speaking for the god, would answer the visitor's question.

Was he a demigod; will he gain immortality? Alexander wanted to know. The Sibyl's response—as always—was laconic, a riddle subject to interpretation. What was clear, though, was the indication that Alexander would find the answer in Egypt—at that country's most famous oracle site: The oasis of Siwa ('D' on map, Fig 1).

* * *

The suggestion was not as odd as it may seem. The two oracle sites were linked by legend and history. The one at Delphi—a name that meant "womb" in Greek—was said to have been chosen by Zeus, head of the Greek pantheon, after two birds he had sent from two opposite places on Earth met there. Declaring the site to be a "navel of the Earth," Zeus placed there an oval-shaped stone called an Omphalus—Greek for "navel." It was a Whispering Stone by which the gods communicated, and according to ancient traditions it was the most sacred object in Apollo's temple, and the Delphic Sybil sat on it when she pronounced her oracular responses. (That original Omphalus stone was replaced in Roman times by a replica, Fig. 3a, which visitors to Delphi can still see.)

The oracle site of Siwa—an oasis in the Western Desert some three hundred miles west of the Nile delta—was likewise chosen after a flight by two black birds (believed to have been priestesses of the god Amon in disguise). The main temple there was dedicated to the Egyptian god Amon, whom the Greeks considered to be the Egyptian 'Zeus'. It too had a Whispering Stone, an Egyptian omphalus (Fig. 3b); and it assumed a sacred place in Greek myth-cum-history because the god Dionysus, once lost in the Western Desert, was saved by being miraculously guided to the oasis. Dionysus was a half-brother of Apollo, and used to fill in for him in Delphi when Apollo was away. Moreover—especially from Alexander's viewpoint—Dionysus attained the status

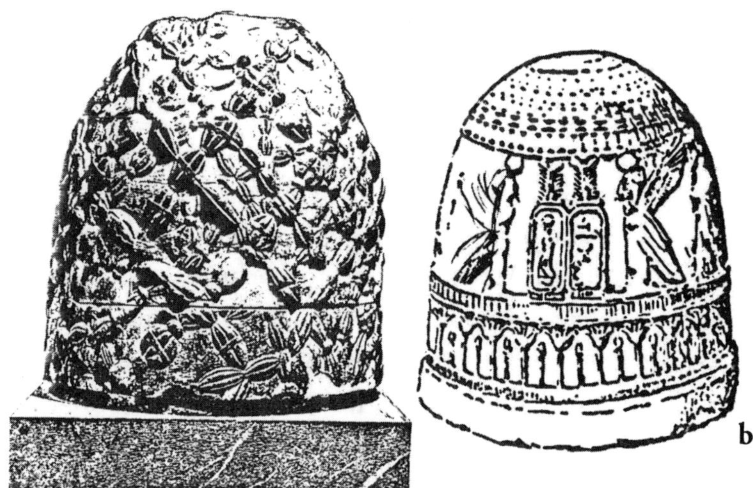

Figure 3

of a god although he was in reality a demigod—the son of Zeus who, disguised as a man, had seduced a princess named Selene. It was, in essence, an earlier occurrence akin to Alexander's—a god in disguise fathering a son by a royal human female; and if Dionysus could be deified and become one of the Immortals—why not Alexander?

Previous seekers of Siwa's oracular pronouncements included two famous generals, Cimon of Athens and Lysander of Sparta; even more significant for Alexander was the demigod Perseus, another love child of Zeus, who managed to slay the monstrous Meduza without turning into stone. The legendary hero Hercules, famed for his challenging Twelve Labors, was also said to have consulted the Siwa oracle; not surprisingly, he too was a demigod, son of Zeus who had impregnated the wise and beautiful Alcmena, having disguised himself as her husband, the king of an island. The precedents clearly fitted Alexander's own quest.

And so it was that instead of pursuing the Persian king and his disarrayed army, Alexander set his course southward. Leaving behind some troops to garrison conquered territory, he marched along the Mediterranean Sea's coastal region. Except for the Phoenician stronghold of Tyre, whose navy participated in the war as Persia's allies, the

Greeks' advance was hardly resisted: Alexander, by and large, was welcomed as a liberator from a detested Persian rule.

In Egypt, the Persian garrison surrendered without a fight, and Alexander was given more than a liberator's welcome by the Egyptians themselves. In Memphis, the capital, the Egyptian priests were ready to accept Alexander's rumored divine parentage by the *Egyptian* god Amon, and they suggested that Alexander travel to Thebes (today's Karnak and Luxor) in Upper Egypt, the site of ancient Egypt's immense temple of Amon, to pay homage to the god and to be crowned a Pharaoh. But Alexander insisted on fulfilling the directive of the Delphi oracle and embarked on the dangerous three-week desert trek to Siwa: He needed to hear the verdict about his immortality.

What had transpired at Siwa during the strictly private oracular session, no one really knows. One version is that when it was over, Alexander said to his companions that he "received the answer that his heart desired," and that "he had learned secret things that he would have not known otherwise." Another version reported that his divine parentage, though not a physical immortality, was confirmed—leading Alexander to henceforth pay his troops with silver coins bearing his image with horns (Fig. 4a), in the likeness of the horned god Amon (Fig. 4b). A third version, supported by what Alexander did there-

Figure 4

after, is that he was instructed to seek out a certain mountain with subterranean passageways in the Sinai Peninsula for angelic encounters, and then proceed to Babylon to the temple of the Babylonian god Marduk.

That latter instruction probably stemmed from one of the "secret things" that Alexander had learned in Siwa: That *Amon* was an epithet meaning "The Unseen" that had been applied in Egypt to the great god **Ra** since about 2160 B.C., when he left Egypt to seek dominion over the whole Earth; his full Egyptian name was **Ra-Amon** or **Amon-Ra**, "the Unseen Ra." In my previous books I have shown that 'Ra-Amon' established his new headquarters in Babylon in Mesopotamia—where he was known as **Marduk**, son of the olden god whom the Egyptians called **Ptah** and the Mesopotamians **Enki**. The secret presumably revealed to Alexander was that his true father, the Unseen (= *Amon*) god in Egypt, was the god Marduk in Babylon; for within weeks of learning all that, Alexander set out for distant Babylon.

As summer began in 331 B.C., Alexander reassembled a large army, and marched toward the Euphrates River on whose banks, south of midstream, Babylon was situated. The Persians, still led by Darius, also assembled a great cavalry and chariot force and awaited Alexander, expecting him to take the traditional route southward along the Euphrates River.

In a great outflanking maneuver, Alexander swerved instead eastward, toward the Tigris River, outflanking the Persians and reaching Mesopotamia in what had historically been Assyria. Learning of Alexander's strategy, Darius rushed troops to the northeast. The two armies met on the eastern side of the Tigris River, at a place called Guacamole ('E' on map, Fig. 1), near the ruins of the erstwhile Assyrian capital Nineveh (now in the Kurdish part of northern Iraq).

Alexander's victory there enabled him to recross the Tigris River; without need to cross the wide Euphrates River, an open plain led to Babylon. Rejecting a third peace offer from Darius, Alexander marched on to Babylon; he reached the renowned city in the autumn of 331 B.C. and rode in through its magnificent Ishtar Gate (reconstruction, Fig. 5;

Figure 5

having been excavated and reassembled, it is now on display in Berlin's Museum of the Ancient Near East).

The Babylonian noblemen and priests welcomed Alexander, delighted to get out of the sway of the Persians who had defiled and demolished Marduk's great temple. The temple was a great ziggurat (step pyramid) at the center of Babylon's Sacred Precinct, rising in seven precise astronomically defined stages (a reconstruction, Fig. 6). Wisely, Alexander let it be known ahead of time that he is coming to pay homage to Babylon's national god, Marduk, and to restore Marduk's defiled temple. It had been a tradition for new Babylonian kings to seek legitimacy by having the deity bless them by grasping their extended hands. But this Alexander could not attain, for he found the god lying dead in a golden coffin, his body immersed in special oils for preservation.

Though surely aware that Marduk had been dead, the sight must

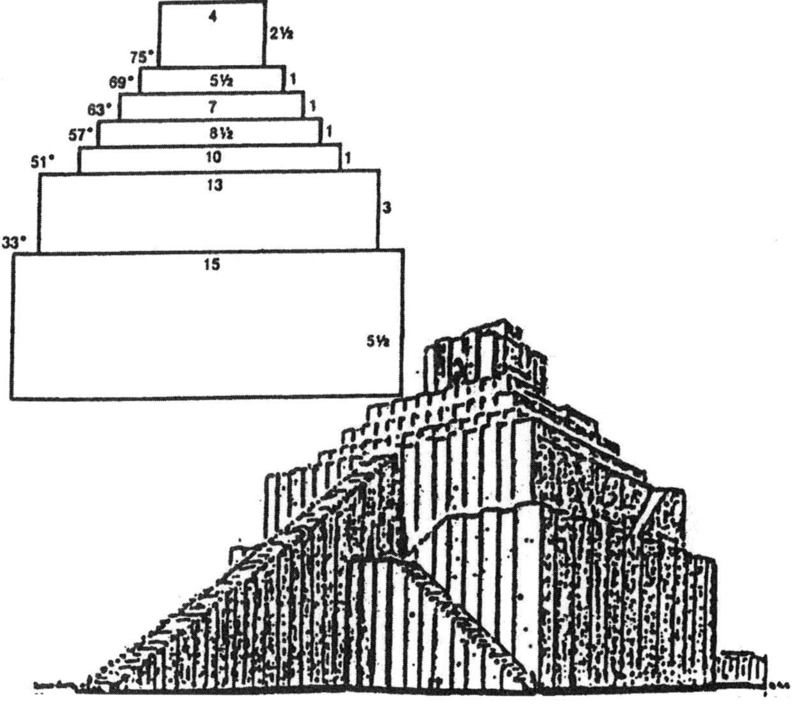

Figure 6

have shocked Alexander: Here lay dead not a mortal, and not just his rumored father, but a *god*—one of the venerated "Immortals." What chance, then, did he, Alexander, a demigod at best, have to avoid death? As if determined to defy the odds, Alexander enlisted thousands of workmen to restore the Esagil, spending scarce resources on the task; and as he left to continue his conquests, he made it clear that he had decided to make Babylon the capital of his new empire.

In 323 B.C., Alexander—by then master of the Persian empire from Egypt to India—returned to Babylon; but Babylonian omen priests warned him not to reenter the city, for he would die if he did. Bad omens, which occurred soon after Alexander's first stay in Babylon, continued although Alexander held off entering the city proper this time. He soon fell ill, seized with high fever. He asked his officers to keep vigil on his behalf inside the Esagil. By the morning of what we now date as June 10, 323 B.C., Alexander was dead—attaining immortality not physically, but by being remembered ever since.

* * *

The tale of Alexander the Great's birth, life, and death has been the subject of books, studies, movies, college courses, and whatever else for generations. Modern scholars do not doubt the existence of Alexander the Great, and have written endlessly about him and his times, ascertaining every detail thereof. They know that the great Greek philosopher Aristotle was Alexander's teacher and mentor, have established Alexander's route, analyzed the strategy of every battle, recorded the names of his generals. But that respected scholars engage in that without an ounce of shame is amazing; for while they describe every aspect and twist in the Macedonian court and its intrigues, they laugh off the part that triggered it all—that of the belief in that court, by Alexander himself, by learned people in Greece—that a god could father a son by a female mortal!

This disdain for "myth" extends to the wider subject of Greek Art. Volumes that make private and public book shelves buckle deal with every minutiae of 'Greek Art' in its varied styles, cultural backgrounds, geographic origins; museums fill up galleries with marble sculptures, bronzes, painted vases, or other artifacts. And what do all of them depict? Invariably—anthropomorphic gods, heroic demigods, and episodes from the so-called mythical tales (as this depiction of the god Apollo welcoming his father, the god Zeus, accompanied by other gods and goddesses, Fig. 7).

For reasons that defy understanding, it is the norm in scholarly circles to classify the records of ancient civilizations thus: If the ancient tale or text deals with kings, it is considered part of Royal Annals. If it deals with heroic personalities, it is an epic. But if the subject is gods, it is classified as Myth; for who in his right scientific mind would believe, as the ancient Greeks (or Egyptians or Babylonians) did, that the gods were actual beings—omnipotent, sky-roaming, engaged in battles, scheming trials and tribulations for heroes—and even fathering those heroes by having sex with human females?

Figure 7

So it is ironic that the saga of Alexander the Great is treated as historical fact, although his birth, oracular visits, itineraries, and end in Babylon could not have taken place without including such 'mythical' gods as Amon, Ra, Apollo, Zeus, and Marduk, or such demigods as Dionysus, Perseus, Hercules—and possibly Alexander himself.

We now know that the lores of all ancient peoples were replete with tales—and depictions—of gods who, though they looked like us, were different—even seemingly immortal. The tales were essentially the same all over the globe; and though the revered beings were named differently in different lands, the names in the diverse languages had by and large the same meaning: an epithet denoting a particular aspect of the named deity.

Thus, the Roman gods called Jupiter and Neptune were the earlier Greek gods Zeus and Poseidon. Indra, the great Hindu god of storms, attained supremacy by battling rival gods with exploding thunderbolts, just as Zeus had done (Fig. 8); and his name, spelled syllabically In-da-ra, was found in god lists of the Hittites in Asia Minor; it was another name for the Hittite chief deity, ***Teshub,*** the god of thunders and lightnings (Fig. 9a)—**Adad** ("Wind Stormer") to the Assyrian and Babylonians,

Figure 8

a

b

Figure 9

Hadad to the Canaanites, and even in the Americas where, as the god ***Viracocha,*** he was depicted on the "Gate of the Sun" in Tiwanaku, Bolivia (Fig. 9b). It is a list that can go on and on. How could that be, why was it so?

Advancing through Asia Minor the Greeks passed imposing Hittite monuments; in northern Mesopotamia they came across the ruins of Assyria's great cities—desolate, but not yet buried by the sands of time. Everywhere, not only the deities names, but also the iconography, the symbols, were the same—dominated by the sign of the Winged Disk (Fig. 10), which they encountered in Egypt and everywhere else—even on the monuments of Persian kings as their supreme symbol. What did it represent, what did it all mean?

Soon after Alexander's death, the conquered lands were split between two of his generals, for his rightful heirs—his four-year-old son and his guardian, Alexander's brother—were murdered. Ptolemy and his successors, headquartered in Egypt, seized the African domains;

Figure 10

Seleucus and his successors, based in Syria, ruled Anatolia, Mesopotamia, and the distant Asian lands. Both new rulers embarked on efforts to learn the full story of the gods and lands now under their control. The Ptolemies, who also established the famed Library of Alexandria, chose an Egyptian priest, known as Manetho, to write down in Greek Egypt's dynastic history and divine prehistory. The Seleucids retained a Greek-speaking Babylonian priest, known as 'Berossus', to compile for them the history and prehistory of Mankind and its gods according to Mesopotamian knowledge. In both instances, the motives were more than mere curiosity; as later events showed, the new rulers sought acceptance by suggesting that their reigns were a legitimate continuation of dynastic kingships that stretched all the way back to the gods.

What we know of the writings of those two savants transports us to the very prehistoric times and events of the intriguing Genesis 6 verses; it takes us beyond the issue of whether "myths" might somehow be true—a collective memory of past events—and catapults us to the discovery that they are versions of actual records, some of which are purported to be *from Days Before The Flood.*

BABYLON AND MARDUK

Called *Bab-Ili* (= 'Gateway of the gods') in Akkadian (from which *Babel* in the Bible), it was the capital city that gave its name to a kingdom on the Euphrates River, north of Sumer & Akkad. Until archaeological excavations begun before World War I brought to light its location and imperial extent, its existence was known only from the Bible—first from the biblical tale of the Tower of Babel, then from historical events recorded in the books of Kings and the Prophets.

The rise and history of Babylon were closely interwined with the fortunes and ambitions of the god **Marduk**, whose main temple—a ziggurat called the *E.sag.il* (= 'House Whose Head is Lofty')—rose within a sprawling sacred precinct, where a plethora of priests hierarchically arranged ranged from cleaners and butchers and healers to administrators, scribes, astronomers, and astrologers. *Mar.duk* (= 'Son of the Pure Mound') was the firstborn son of the Sumerian god Ea/Enki, whose domains were in Africa (where, I have suggested, they were worshipped as the gods Ra and Ptah, respectively). But Marduk sought overall dominion by establishing his own 'Navel of the Earth' in Mesopotamia proper—an effort that included the failed 'Tower of Babel' incident. Success finally came after 2000 B.C., when a resplendent Marduk (see illustration, next page) invited all the other leading gods to reside in Babylon as his subordinates.

Babylonia attained imperial status with the dynasty started by King Hammurabi circa 1800 B.C. The deciphering of cuneiform texts found all over the ancient Near East provided historical data about its religion-motivated conquests and rivalry with Assyria. After a decline that lasted some five centuries, a Neo-Babylonian empire rose again, lasting to the 6th century B.C. Its conquests included several attacks on Jerusalem and the destruction of its Temple in 587 B.C. by King Nebuchadnezzar II—fully corroborating the biblical tales.

The city of Babylon, as an imperial capital, a religious center, and the symbol of its kingdom, came to an end in 539 B.C. with its capture by the Achaemenid-Persian King Cyrus. While he was respectful of Marduk, his successor, Xerxes, destroyed the famed ziggurat-temple in 482 B.C., for by then it only served as a glorified tomb for the dead Marduk. It was those ruins of the ziggurat-temple that Alexander attempted to rebuild.

II
In the Days before the Flood

Enlisted by King Ptolemy Philadelphus circa 270 B.C., Manetho (Greek from Men-Thoth = 'Gift of Thoth') compiled the history and prehistory of ancient Egypt in three volumes. The original manuscript, known as Aegyptiaca, was deposited in the Library of Alexandria, only to perish there with other irreplaceable literary and documentary treasures in natural and man-made calamities, including its final burning by Moslem conquerors in A.D. 642. We do know, however, from quotes and references in the writings of others in antiquity (including the Jewish-Roman historian Josephus) that Manetho listed gods and demigods as reigning long before human Pharaohs became kings in Egypt.

The Greeks were not completely ignorant of Egypt and its past, certainly since the historian-cum-explorer Herodotus visited the land two centuries earlier. On the subject of Egypt's rulers, Herodotus wrote that Egyptian priests "said that Mên was the first king of Egypt." True to apparently the same sources, Manetho's list of Pharaohs also began with one called Mên (Menes in Greek); but it was Manetho who was first to arrange the succession of Pharaohs by dynasties—an arrangement followed to this day—combining genealogical affiliations with historical changes. His comprehensive List of Kings gave their names, lengths of reign, order of succession, and some other pertinent information.

What is significant in Manetho's list of Pharaohs and their dynasties is that *his list begins with gods* and not with Pharaohs. Gods and demigods, Manetho wrote, reigned over Egypt before any human Pharaohs did!

Their names, order, and lengths of reigns—"fabulous," "fantastic" scholars say—began with a divine dynasty headed by the god **Ptah**, ancient Egypt's Creator God:

Ptah	reigned	9,000 years
Ra	reigned	1,000 years
Shu	reigned	700 years
Geb	reigned	500 years
Osiris	reigned	450 years
Seth	reigned	350 years
Horus	reigned	300 years
Seven gods	reigned	12,300 years

Like his father Ptah, **Ra** was a god "of Heaven and Earth," having arrived in earlier times from the "Planet of Millions of Years" in a Celestial Barque called the *Ben-Ben* (meaning 'Pyramidion Bird'); it was kept in the Holy of Holies of a shrine in the sacred city Anu (the biblical *On,* better known by its later Greek name Heliopolis). Though benefiting from unbridled longevity and playing a role in Egyptian affairs for millennia to come, Ra's reign as Ptah's successor was cut short—abruptly—after a mere one thousand years. The reason, we shall find, was significant to our quest.

The first divine dynasty that ended with Horus, Manetho reported, was followed by a second one, headed by the god **Thoth** (another son of Ptah, but only a half-brother of Ra). Its reign lasted a total of 1,570 years. In all, Manetho said, gods ruled for 13,870 years. A dynasty of thirty ***demigods*** followed; they reigned a total of 3,650 years. All in all, Manetho wrote, divine and semi-divine rulers reigned a total of 17,520 years. Then, after a chaotic intermediate period that lasted 350 years, with no one reigning over the whole (i.e., both Lower and Upper) of Egypt, *Mên* began the first human dynasty of Pharaohs, ruling over a unified Egypt.

Various modern archaeological discoveries that corroborate Manetho's Pharaohnic list and order of succession include a document known as the Turin Papyrus and an artifact called the Palermo Stone, so named after the museums in Italy in which they are kept. The corroborating finds also include a stone inscription known as the Abydos List, in which the 19th dynasty Pharaohs Seti I and his son Ramses II, who reigned a thousand years before Manetho's time, depict themselves (Fig. 11). Carved on the walls of the main temple in Abydos, a city in Upper Egypt, it lists the names of seventy-five of their predecessors, beginning with "Mêna." The Turin Papyrus corroborates Manetho's divine, semidivine, and chaotic-interval lists, and (including subsequent Pharaohs) names a total of 330 rulers, just as Herodotus had been told.

The famed Egyptologist Sir W. M. Flinders Petrie excavated a group of tombs in a most ancient cemetery on the outskirts of Abydos. Stelas that served as tombstones and other inscriptions identified the place—located beside a purported Tomb of Osiris—as the burial grounds of First and Second dynasty Pharaohs; the sequence of tombs, from east to west, began with one bearing the name of King Menes. Petrie identified tombs bearing the names of all of the First Dynasty Pharaohs, and in his masterwork, *The Royal Tombs of the First Dynasty* (1900/1901), acknowledged that the finds confirmed Manetho's list. Moreover,

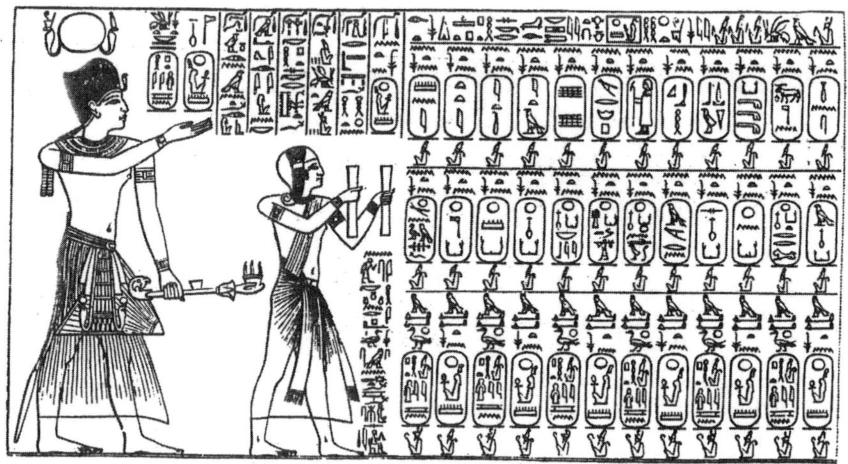

Figure 11

he found tombs with names of pre-dynastic kings, nicknaming them "Dynasty 0." Subsequent Egyptologists have identified them as rulers during the Chaotic Period listed by Manetho, corroborating that part of his listings too.

The importance of such corroborated data goes beyond the issue of divine and semi-divine dynasties in pre-Pharaonic times: It throws significant light on the subject of the Deluge and pre-diluvial times. Since it is now known with certainty that Pharaonic rule began in Egypt circa 3100 B.C., Manetho's timeline takes us back to 20,970 B.C. (12,300 + 1,570 + 3,650 + 350 + 3,100 = 20,970). Climate and other data presented in my books *The 12th Planet* and *Genesis Revisited* led to the conclusion that the Deluge occurred some 13,000 years ago, circa 10,970 B.C.

The resulting ***difference of 10,000 years (20,970–10,970) is exactly the length of the combined divine reign of Ptah (9,000 years) and the abruptly cut short reign of Ra (1,000 years).*** This is a significant synchronism that links the Manetho timetable to the Deluge. It suggests that Ptah reigned before the Deluge, and that Ra's reign was abruptly cut short by the Deluge. It confirms the reality of the Deluge and its timing on the one hand, and the veracity of Manetho's divine and demigod data on the other hand.

Astounding as this synchronism is, it is not merely coincidental. The Egyptians called their country "The Raised Land," because at one time, ancient lore said, it was inundated by an engulfing avalanche of water that completely flooded the land. The god Ptah, a great scientist, came to the rescue. On the Nile River's island Abu (also called Elephantine for its shape), near the river's first cataract in Upper Egypt, Ptah formed a cavern in the mighty rocks and installed in it sluices that controlled the river's flow, enabling the ground downstream to dry out—literally, in Egyptian eyes, raising the land from under the waters. The feat was depicted in Egyptian art (Fig. 12); the modern great dam at Aswan is located at the same site near the first cataract.

These events may well offer an explanation why the god who then assumed reign over Egypt was called *Shu,* whose name—'Dryness'—bespoke the end the watery catastrophe. His successor bore the name

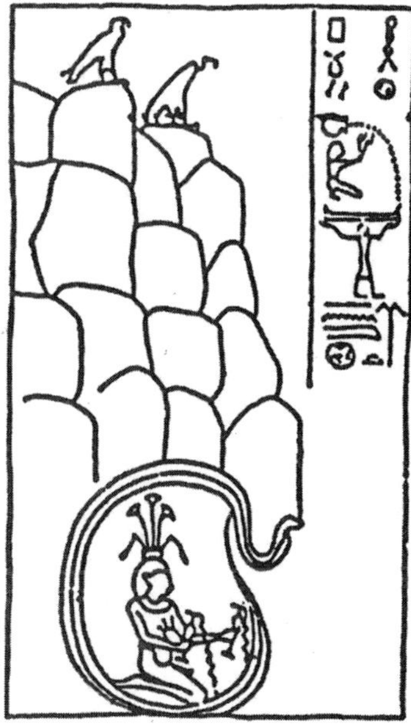

Figure 12

Geb (meaning 'He who heaps up'), for he engaged in great earthworks to make the land even more habitable and productive. Like pieces in a jigsaw puzzle, all of these diverse facts add up to an Egyptian record of a Deluge, a Great Flood, circa 10,970 B.C.

To these tidbits of Egyptian prehistory concerning the Deluge, one can add the fact that as a way to unify Egypt, Mên emulated Ptah by creating an artificial island in the Nile, where it begins to branch out into a delta, and built there a new capital dedicated to the god Ptah; he named it Mên-Nefer ('The Good Place of Mên')—Memphis in Greek.

Like Greek history and art, the history and prehistory of ancient Egypt cannot be divorced from the active presence and physical existence of its gods. Wherever one looks or turns in Egypt, the statues, sculptures, depictions, temples, monuments, texts inscribed and illustrated inside pyramids or on coffin lids or on the walls of tombs—all speak of, name, and depict Egypt's gods and its leading pantheon (Fig. 13). Whatever had been recorded and depicted before Manetho's days, and discovered after his time, corroborates his lists of Pharaonic dynasties; *why not also*

THE CELESTIAL DISK AND THE GODS OF EGYPT

1. Ptah	2. Ra-Amen	3. Thoth	4. Seker
5. Osiris	6. Isis with Horus	7. Nephtys	8. Hathor

The gods with their attributes:

9. Ra/Falcon 10. Horus/Falcon 11. Seth/Sinai Ass 12. Thoth/Ibis 13. Hathor/Cow

Figure 13

accept the reality of gods, followed by demigods, as rulers in Egypt preceding human Pharaohs?

* * *

In the Seleucid domains, the task of compiling the tale of the past was assigned to a Babylonian priest-historian named Berossus (Greek from *Bel-Re'ushu* = 'The Lord [Bel = Marduk] is his shepherd') who was born in Babylon when Alexander the Great was there. His task was much more complex than that of Manetho in Egypt, for his compilation was not limited to one land; it had to embrace many lands, different kingdoms, and diverse rulers who reigned not necessarily in succession but sometimes contemporaneously in different (and sometimes warring) capitals.

The three volumes he had composed (called *Babyloniaca* and dedicated to King Antiochus I, 279–261 B.C.) are no longer extant, but portions of them were retained, having been copied and extensively quoted in antiquity by contemporary Greek savants, and later on by other Greek and Roman historians (including Josephus). It is from those references and quotes, collectively known as "Fragments of Berossus," that we know that Berossus chose to 'globalize' the subject: He chose to write down not the history of one nation or one kingship, but of the whole Earth, not of one group of gods but of all the gods, of Mankind in general, of how it all—gods, demigods, kingship, kings, human beings, civilization—had come to be; a comprehensive history from The Beginning to Alexander's time. It is from those Fragments that we know that Berossus divided the past into a time before a Great Flood and the eras after the Flood, and asserted that before there were men, gods alone ruled the Earth.

Alexander Polyhistor, a Greek-Roman historian-geographer in the 1st century B.C., reported in regard to the pre-Diluvial era that "in the second book [of Berossus] was the history of the ten kings of the Chaldeans, and the periods of each reign, which consisted collectively of a hundred and twenty *Shars,* or four hundred and thirty-two thousand years, reaching to the time of the Deluge." ('Chaldeans' was a term used to describe the astronomically savvy residents of ancient Mesopotamia.)

The grand total of 432,000 years comprised the combined reigns of the ten listed rulers, whose individual reigns lasted anywhere from 10,800 to 64,800 years. The Greek historians who quoted Berossus explained that the great lengths of those rulers' reigns were actually given in number units called *Shar,* each *Shar—Saros* in Greek—being equal to 3,600 years. The Greek historian Abydenus, a disciple of Aristotle, who quoted Berossus, made clear that these ten rulers and their cities were all in ancient Mesopotamia and explained how their reign periods were rendered:

> It is said that the first king of the Earth was Aloros;
> he reigned ten *Shars.* Now, a *Shar* is esteemed to be three thousand six hundred years.
> After him Alaprus reigned three *Shars.*
> To him succeeded Amillarus from the city of panti-Biblon, who reigned thirteen *Shars.*
> After him Ammenon reigned twelve *Shars;* he was of the city of panti-Biblon.
> Then Megalurus of the same place, eighteen *Shars.*
> Then Daos, the Shepherd, governed for the space of ten *Shars.*
> Afterward reigned Anodaphus and Euedoreschus.
> There were afterward other rulers, and the last of all Sisithrus;
> so that in the whole their number amounted to ten kings, and the term of their reigns to a hundred and twenty *Shars.*

Apollodorus of Athens (2nd century B.C.) also reported on the pre-Diluvial disclosures by Berossus in similar terms: Ten rulers reigned a total of 120 *Shars* (= 432,000 years), and the reign of each one of them was measured in the 3,600-year *Shar* units. Indeed, all those who had quoted Berossus affirmed that he listed ten divine rulers who had reigned from the beginning until the Great Flood, treating the Deluge as a decisive event. The names of the ten pre-Diluvial rulers (rendered as Greek names by those who quoted Berossus) and the lengths of their reigns, totaling 120 *Shars,* were as shown on page 29. (Though the sequences of succession varied, all quotings agree that an "Aloros" was the first and a "Xisuthros" the last.)

Aloros	reigned for	10 Shars	(= 36,000 years)
Alaparos	reigned for	3 Shars	(= 10,800 years)
Amelon	reigned for	13 Shars	(= 46,800 years)
Ammenon	reigned for	12 Shars	(= 43,200 years)
Megalarus	reigned for	18 Shars	(= 64,800 years)
Daonos	reigned for	10 Shars	(= 36,000 years)
Euedoreschus	reigned for	18 Shars	(= 64,800 years)
Amempsinos	reigned for	10 Shars	(= 36,000 years)
Obartes	reigned for	8 Shars	(= 28,800 years)
Xisuthros	reigned for	18 Shars	(= 64,800 years)
Ten rulers	reigned for	120 Shars	(= 432,000 years)

The quotings from Berossus indicate that his writings dealt with several issues concerning Mankind itself—how it came to be, how it attained knowledge, how it spread and settled the Earth. In the beginning gods alone were upon the Earth. Men appeared, according to the Berossus Fragments, when *Deus* ("god"), also called *Belos* (a name meaning 'Lord'), decided to create Man. He used for the purpose a "twofold principle," but the results were "hideous Beings." "Men appeared with two wings, some with four, and with two faces . . . Other human figures were to be seen with the legs and horns of goats . . . Bulls likewise bred there with the heads of men . . . Of all these were preserved delineations in the temple of Belus in Babylon." (*Belus,* Greek for *Bel/Ba'al,* "the Lord," was in Babylon an epithet for the god Marduk).

On the subject of how men attained intelligence and knowledge, Berossus wrote that it came about thus: A leader of those early divine rulers named ***Oannes*** waded ashore from the sea and taught Mankind all aspects of civilization. "He was a Being endowed with reason, a god who made his appearance from the Erythrean Sea that bordered on Babylonia." Berossus reported that although Oannes looked like a fish, he had a human head under the fish's head, and had feet like a man under the fish's tail. "His voice too and language were articulate and human." ("A representation of him," Alexander Polyhistor added, "is preserved even to this day.")

This Oannes "used to converse with men; he gave them insight into letters and sciences and every kind of art; he taught them to construct houses, to found temples, to compile laws; and explained to them the principles of geometrical knowledge." It was Oannes, according to the Fragments recorded by Polyhistor, who put in writing a tale that explained how Mankind came to be, Creation having been preceded by "a time in which there was nothing but darkness and an abyss of waters."

The Berossus Fragments then include details regarding the defining event, the Great Flood, that separated the era of the gods from the times of men. According to Abydenus, Berossus reported that the gods kept knowledge of the coming devastating Deluge a secret from Mankind; but the god Cronus (in Greek legends, a son of the god Uranus = Sky and the father of the god Zeus) revealed the secret to "Sisithros" (= the last-named ***Xisuthros*** of the ten pre-Diluvial rulers):

> Cronus revealed to Sisithros that there would be a
> Deluge on the fifteenth day of Daisios, and ordered
> him to conceal in Sippar, the city of the god Shamash,
> every available writing.
> Sisithros accomplished all these things, and sailed
> immediately to Armenia; and thereupon what the god
> had announced did happen.

To find out whether the Deluge had ended, according to the Abydenus quotes, Sisithros released birds to see if they would find dry land. When the boat reached Armenia, Sisithros made sacrifices to the gods. He instructed the people who were with him in the boat to go back to Babylonia; as for himself, he was taken by the gods to spend the rest of his life with them.

Polyhistor's account was longer and more detailed. After reporting that "after the death of Ardates [or Obartes] his son Xisuthros ruled for eighteen Sars and in his time the Great Flood occurred," Polyhistor rendered the Chaldean account of it thus:

> The deity, Cronus, appeared to him in a vision and
> gave him notice that on the fifteenth day of the month
> Daisos there would be a Flood by which Mankind
> would be destroyed.
> He enjoyned him to commit to writing a history of the
> Beginnings, Middles, and Ends of all things, down to
> the present term; and to bury those accounts securely in
> the city of the Sun god, in Sippar;
> And to build a vessel, and to take into it with him
> his kinsfolk and his friends
> He was to stow food and water and put birds and animals
> on board, and sail away when he had everything ready.

Following these instructions Xisuthros built a boat, "five stades long and two stades wide." Anticipating some raised eyebrows from the other townspeople, Xisuthros was instructed by his god to just say he was "sailing to the gods, to pray for blessings on men." He then put on board his wife and children "and closest friends."

When the Flood subsided, "Xisuthros let out some of the birds, which finding no food came back to the vessel." On the third try, the birds did not return and Xisuthros inferred that land had appeared. After the boat ran aground, Xisuthros, his wife, his daughter, and his pilot went ashore, and were never seen again, "for they were taken to dwell with the gods." Those who were left behind on board were told by an unseen voice that they were in Armenia and were instructed to return to their land and "rescue the writings from Sippar and disseminate them to Mankind." This they did:

> They came back to Babylon, they dug up the writings
> from Sippar, founded many cities, set up shrines, and
> once again established Babylon.

According to the Fragments, Berossus wrote that at first "all men spoke the same language." But then "some among them undertook to erect a large and lofty tower, that they might climb up to heaven." But

Belus, sending forth a whirlwind, "confounded their designs and gave each tribe a particular language of its own." "The place at which they built the tower is now called Babylon."

* * *

The similarities between the Berossus tales and those in the Bible's book of *Genesis* are readily obvious; they extend beyond the subject of the Deluge and match each other in many details.

The Deluge, according to Berossus, occurred in the reign of the 10th pre-Diluvial ruler, Sisithros, and began in the month Daisos, which was the second month of the year. The Bible (Genesis 7:12) likewise states that the Deluge occurred "in the six hundredth year of Noah's life, *in the second month,*" Noah having been the 10th pre-Diluvial biblical Patriarch (starting with Adam).

Like Xisuthros/Sisithros, Noah was told by his god that a devastating avalanche of water is about to happen, and was instructed to build a waterproof vessel according to precise specifications. He was to take aboard his family, animals, and birds—as Xisuthros did. When the waters subsided, both released birds to see if dry land reappeared (Noah sent two birds, first a raven, then a dove). Sisisthros's boat came to rest "in Armenia"; Noah's ark came to rest in the "mountains of Ararat," which are in Armenia.

Another major event is similarly reported by both the Bible and Berossus: The incident of the Tower of Babel that resulted in the Confusion of Languages. We have quoted above the Berossus version; like it, the Bible begins the tale (in Genesis 11) with the statement that at that time "The whole Earth was of one language and one kind of words." Then the people said, "let us build a city and a tower whose top can reach the heavens"; Berossus states the same thing: People set out "to erect a large and lofty tower, that they might climb up to heaven." In the Bible, God ("Yahweh") "came down to see the city and the tower that the Children of the Adam had built." He got concerned and "confused their language so that they may not understand each other" and "scattered the people over the face of the Earth." Berossus ascribes the Confounding of Languages to The Lord ("Belus") and attributes Mankind's scattering to the deity's use of a Whirlwind.

Do such similarities mean that the opening chapters of Genesis are one large 'Fragment of Berossus', that the compilers of the Hebrew Bible copied from Berossus? Not likely, for the whole Torah part of the Hebrew Bible, its first five books from Genesis to Deuteronomy, was already "sealed"—canonized in a final version unchanged since then—long before the time of Berossus.

It is a historical fact that the Hebrew Bible was already in its 'sealed' version when the five Torah books and the rest of the Bible were translated in Egypt into Greek by order of the same Ptolemy Philadelphus (285–244 B.C.) who had engaged Manetho to write Egypt's history. The translation, still extant and available, is known as the *Septuagint* ("Of the Seventy") because it was carried out by a group of seventy scholars. A comparison of its Greek text with the Hebrew Bible leaves no doubt that those savants had already in front of them the canonized version of the Hebrew Bible as we know it today—a Bible that was already in its final form *before* the time of Berossus (and Manetho).

Did then Berossus use the Hebrew Bible as his source? That too is unlikely. Apart from his references to 'pagan' gods (Cronus, Belus, Oannes, Shamash) who are absent in the monotheistic Bible, many particulars in his writings are not found in the biblical version, so his sources had to be other than the Bible. A most significant difference occurs in the Creation of Man tale, with its terrifying mishaps in the Berossus version, in contrast to the smooth "Let us fashion the Adam" version in the Bible.

There are detail differences even where the two versions dovetail, as in the Deluge story in regard to the size of the ship and, more importantly, as to who was taken on board to be saved. Some of the differences are not insignificant: According to Berossus, there were on board, beside the immediate family of 'Noah', also several of his friends, as well as a skilled pilot; not so in the Bible that listed just Noah, his wife, and their three sons and their wives. This is not a minor matter: ***If true,***

then post-Diluvial Mankind, genetically and genealogically, does not stem solely from one Noah and his only three sons.

The whole tale of Oannes, the god dressed as a fish, wading ashore to grant civilization to Mankind, is nowhere in the Bible. Also absent in the Bible is the reference to a pre-Diluvial city named Sippar ("the city of the Sun-god Shamash") and the safekeeping there of "every available writing." By claiming that pre-Diluvial records of "Beginnings, Middles, *and Ends*" not only had once existed but were hidden for safekeeping and were retrieved after "Babylon" was resettled, Berossus could have sought legitimacy for his version of prehistoric events; but he also suggested that those Records of the Past contained clues to the Future—what the Bible, and we nowadays, call 'The End of Days'. Though the theme of linking the Future to the Past is part of biblical prophecy, in the Bible it is first mentioned in respect to Jacob—long after the Deluge.

The logical conclusion—that both the compilers of Genesis and then Berossus had access to the same or similar source material, which each used selectively—has been borne out by archaeology. But in such a conclusion, both similarities and differences take us back to our starting point, the enigmatic verses in Genesis 6: Who were the *Nefilim,* who were the sons of the gods—***and who, indeed, was Noah***?

THE SHIP OF NOAH

In the Sumerian text, the ship of Ziusudra was termed ***Ma.gur.gur*** = a 'ship that can tumble and turn'. In the Akkaddian texts it was referred to as a *Tebitu,* with a hard 'T', meaning a submersible ship; the biblical redactor rendered it with a soft 'T', a *Teba*—a 'Box' (hence the 'ark' in translations). In all versions, it was hermetically sealed with bitumen but had one openable aperture.

According to the Epic of Gilgamesh, the ship which *Utnapishtim,* the name of the hero of the Deluge in Akkadian, was instructed to build was 300 cubits (about 525 feet) long, 120 cubits (about 210 feet) wide at the top, and had a "bulwark" (height) of 120 cubits divided by 6 decks into 7 levels, "one-third of her above the water line."

Genesis 6:15 also reports a length of 300 cubits, but only 50 cubits (about 88 feet) of width, and only 30 cubits (about 53 ft.) height, with only 3 stories (the roofed upper one included).

At the beginning of the 20th century, biblical scholars drew comparisons with the largest-ever passenger ships then known to them:

The Great Eastern, built in 1858, was 680 ft. long, 83 ft. wide, 48 ft. deep;

The City of Rome, built in 1881, measured 560, 52, and 37 ft., respectively;

The famed *Lusitania,* 1907, measured 762, 88, and 57 ft., respectively;

Her sister ship, *The Mauretania,* was the first to have 8 decks.

Those modern proportions of length/width/height seem to agree

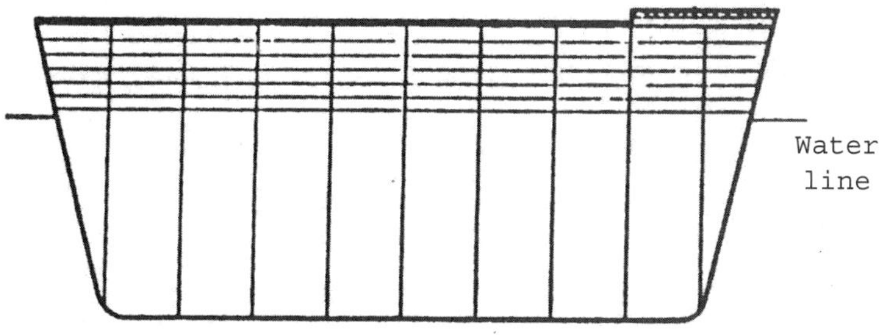

more with the biblical description: Noah's Ark was as long as *The City of Rome,* as wide as *The Great Eastern,* and as high as *The Lusitania.*

In his 1927 study "The Ship of the Babylonian Noah" the Assyriologist Paul Haupt suggested the design shown on page 35, based on the various ancient texts.

III
In Search of Noah

The deciphering of Egyptian hieroglyphic writing was decisively facilitated by the chance discovery, during Napoleon's expedition to Egypt in 1799, of the Rosetta Stone—a stone tablet from 196 B.C. (now on display in the British Museum, Fig. 14) on which a royal Ptolemaic

Figure 14

proclamation was inscribed in three languages: Egyptian hieroglyphics, a later Egyptian cursive script called Demotic, and Greek. It was the Greek part that served as a key to unlocking the secrets of ancient Egypt's language and writing.

No 'Rosetta Stone', a single decisive master discovery of a tablet, had occurred in the ancient Near East; there, the process of discovery was long and tedious. But there too, other forms of multilingual inscriptions moved decipherment along; above all, progress was made when it was realized that the Bible—the Hebrew Bible—was a key for unlocking those enigmatic writings. By the time decipherment was attained, not only several languages but *several ancient empires*—one of them most astounding—came to light.

Fascinated by the tales (magnified as centuries passed) of Alexander the Great and his conquests, European travelers ventured to faraway Persepolis (Greek for 'City of the Persians'), where remains of palaces, gateways, processional stairways, and other monuments were still standing (Fig. 15). Visible engraved lines (that turned out to be inscriptions) were assumed at first to be some form of decorative design. A 1686 visitor (Engelbert Kampfer) to the ruins of this Persian royal site described the marking as "cuneates" ('Wedge-Shaped'—Fig. 16); the designation, 'Cuneiform' has stuck ever since to what has in time been recognized as a lingual script.

Cuneiform script variations on some monuments gave rise to the idea that, as had been the case in Egypt, royal Persian proclamations in an empire that encompassed many diverse peoples could also be multilingual. Disparate reports by travelers increasingly focused attention on some of the multilingual Persian inscriptions; the most important and complex of them was discovered at a site in what is now northern Iran. It was in 1835 that traveling in remote Near Eastern areas that were once dominated by Persian kings, the Briton Henry Rawlinson came across a carving upon forbidding rocks at a place called Behistun. The name meant 'Place of gods', and the huge carving commemorating a royal victory was dominated by a god hovering within the ubiquitous Winged Disc (Fig. 17). The depiction was accompanied by long

Figure 15

inscriptions that (once deciphered by Rawlinson and others) turned out to be a trilingual record by the Persian king Darius I, a predecessor, by a century and a half, of Darius III who fought Alexander.

Figure 16

Figure 17

In time it was realized that one of the Behistun languages, dubbed Old Persian, resembled Sanskrit, the 'Indo-European' mother language; it was a finding that opened the way to the decipherment of Old Persian. Taking it from there, the identity and meaning of the other two languages followed. One was later identified as Elamite, whose use in antiquity was limited to the southern parts of what is now Iran. The third matched the writings found in Babylonia; classified as 'Semitic', it belonged to a group that also included Assyrian and Canaanite whose mother tongue is called 'Akkadian'. What was common to all three Behistun languages was the use of the same cuneiform script, in which each sign expresses a whole syllable and not just a single letter. Here, in one monument, was an example of the Confusion of Languages . . .

Hebrew, the language of the Bible, belonged to the group of 'Semitic' languages that stemmed from 'Akkadian'. The fact that Hebrew, uniquely, has remained a spoken, read, and written language throughout the ages was the unlocking key there—so much so that early scholarly studies of Babylonian and Assyrian (two 'Akkadian' languages) provided word lists that gave their similar Hebrew meanings, and compared cuneiform sign lists to their equivalents in traditional Hebrew script (Fig. 18—from *Assyrian Grammar* by the Rev. A. H. Sayce, 1875).

Word of intriguing ruins in the great plain between the Euphrates and Tigris Rivers (hence *Mesopotamia,* 'Land Between the Rivers') has been brought back to Europe by varied 17th- and 18th-century travelers. Then, suggestions that such ruins represented Babylon and Nineveh of biblical fame (and wrath) stirred up a more active interest. The realization that 19th century A.D. people were able to read inscriptions of people from a time before Greece and Persia, inscriptions from the time of the Bible, shifted interest geographically to the Lands of the Bible and chronologically to much earlier centuries.

In some of those ruins, inscriptions in cuneiform script were found on flat tablets—tablets that were man-made of hardened clay, mostly but not always square or oblong in shape, into which the wedgelike signs

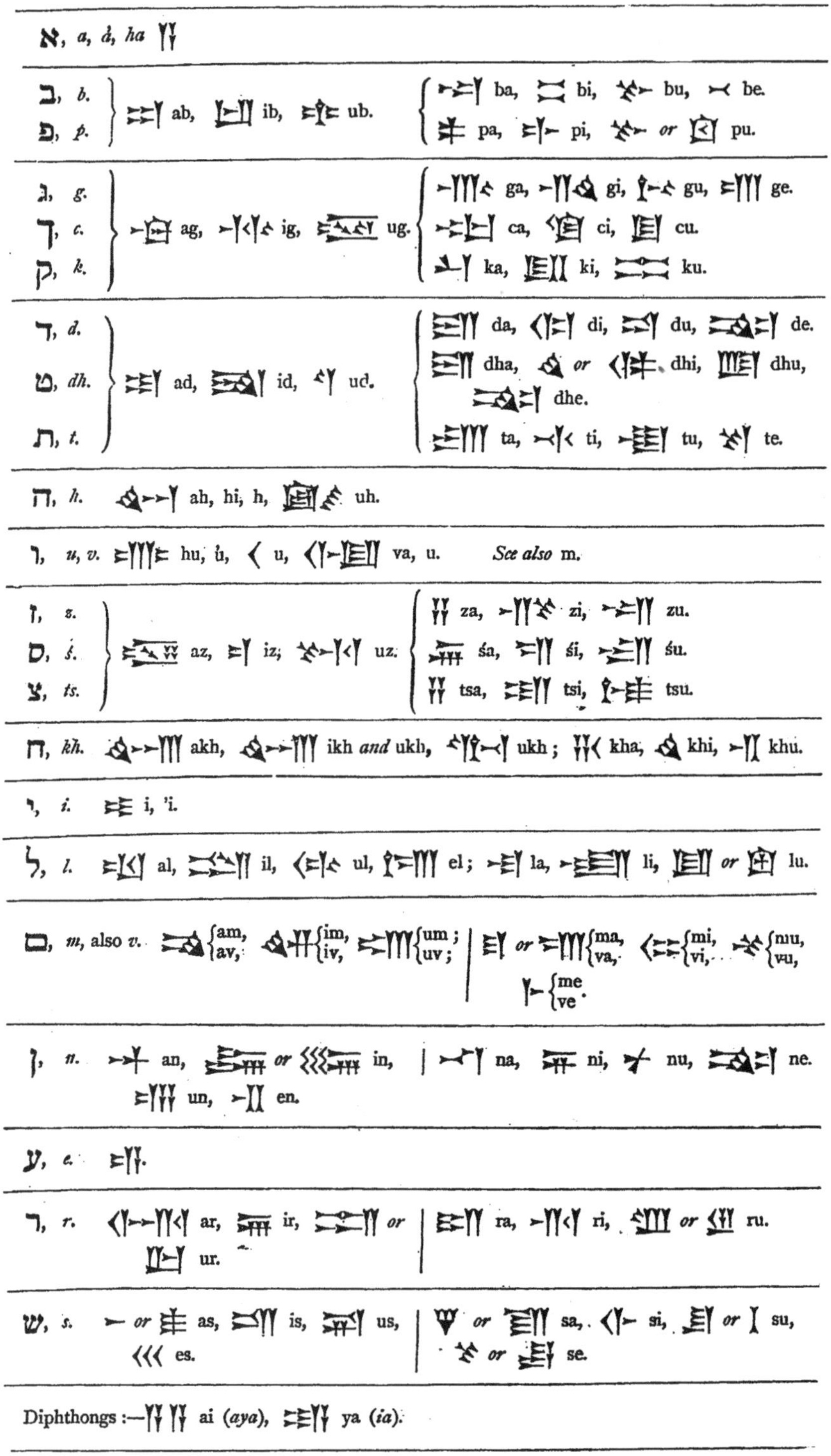

Letter	Value	Syllables (closed)	Syllables (open)
א,	*a*, *â*, *ha*		
ב, פ,	*b.*, *p.*	ab, ib, ub.	ba, bi, bu, be. pa, pi, *or* pu.
ג, ך, ק,	*g.*, *c.*, *k.*	ag, ig, ug.	ga, gi, gu, ge. ca, ci, cu. ka, ki, ku.
ד, ט, ת,	*d.*, *dh.*, *t.*	ad, id, ud.	da, di, du, de. dha, *or* dhi, dhu, dhe. ta, ti, tu, te.
ח,	*h.*	ah, hi, h, uh.	
ו,	*u*, *v.*	hu, ů, u, va, u. *See also* m.	
ז, ס, צ,	*z.*, *ś.*, *ts.*	az, iz, uz.	za, zi, zu. śa, śi, śu. tsa, tsi, tsu.
ח,	*kh.*	akh, ikh *and* ukh, ukh; kha, khi, khu.	
י,	*i.*	i, 'i.	
ל,	*l.*	al, il, ul, el; la, li, *or* lu.	
ם,	*m*, also *v.*	am, av, im, iv, um, uv;	*or* ma, va, mi, vi, mu, vu, me, ve.
ן,	*n.*	an, *or* in, un, en.	na, ni, nu, ne.
ע,	*e.*		
ר,	*r.*	ar, ir, *or* ur.	ra, ri, *or* ru.
ש,	*s.*	*or* as, is, us, es.	*or* sa, si, *or* su, *or* se.

Diphthongs :— ai (*aya*), ya (*ia*).

Figure 18

were incised when the clay was still wet and soft (Fig. 19). Curious what they represented and what they said, European consuls stationed in various parts of the Ottoman empire pioneered what can be considered modern Near Eastern archaeology; its beginning—excavating ancient Babylon—took place south of Baghdad in Iraq in 1811. (In a twist of fate, clay tablets discovered in the ruins of Babylon included several whose cuneiform inscriptions recorded payments in silver coins by Alexander for work done in clearing debris from the Esagil temple.)

In 1843 Paul Emile Botta, the French Consul in Mosul, a town now in Kurdish northern Iraq in what was then Ottoman-ruled Mesopotamia, set out to excavate an ancient source of such clay tablets at a *Tell* (ancient mound) near Mosul. The site was named Kuyunjik after the nearby village; an adjoining Tell was called *Nebi Yunus* ('Prophet Jonah') by the local Arabs. Botta abandoned the site after his initial probes there were

Figure 19

unproductive. Not to be outdone by the French, the Englishman A. Henry Layard took over the site three years later. The two mounds, where Layard was more successful than Botta, proved to be the ancient Assyrian capital Nineveh that is mentioned repeatedly in the Bible, and that was Jonah's destination according to the Bible's tale of Jonah and the Whale.

Botta found success farther north, at a site called Khorsabad, where he uncovered the capital of the Assyrian king Sargon II (721–705 B.C.) and his successor, King Sennacherib (705–681 B.C.); Layard gained fame as the discoverer of both *Nineveh* and, at another site called locally *Nimrud,* of the Assyrian royal city *Kalhu* (named Calah in the Bible). Not counting Babylon, the finds by both provided, for the first time, physical evidence corroborating the Bible (Genesis, chapter 10) about the hero *Nimrod* and Assyria and its major cities:

> He was first to be a hero in the Land;
> And the beginning of his kingdom:
> Babel and Erech and Akkad,
> all in the Land of Shine'ar.
> Out of that Land there emanated *Ashur,*
> where *Nineveh* was built—a city of wide streets,
> and *Calah,* and Ressen—the great city
> which is between *Nineveh* and *Calah.*

At Khorsabad the excavators uncovered, among the lavish wall reliefs glorifying Sennacherib and his conquests, panels depicting his siege of the fortified city of Lachish in Judea (in 701 B.C.). The Bible (2 Kings and in Isaiah) mentions that siege (in which Sennacherib prevailed) as well as his failed siege of Jerusalem. Layard's finds included a stone column of the Assyrian king Shalmaneser III (858–824 B.C.) that described, in text and carved drawing, his capture of King Jehu of Israel (Fig. 20)—an event reported in the Bible (2 Kings, 2 Chronicles).

Wherever finds were made, it seemed, it was like digging up the veracity of the Bible.

(By another twist of fate, Layard's sites Nimrud and Nineveh were on opposite sides of the river bend where Alexander had crossed the

Figure 20

Tigris River and delivered the final blow to the Persian army.)

By the end of the 19th century, as the rumblings of the conflagration known as World War I became more ominous, the Germans joined the archaeological race (with its mapmaking, spying, and influence-peddling ramifications). Outflanking the French and the British, they took control of sites farther south, uncovering at Babylon (under the leadership of Robert Koldewey) most of the sacred precinct, the Esagil temple-ziggurat, and the grand Processional Way with its varied gates including that of Ishtar (see Fig. 5). Farther north Walter Andrae unearthed the olden Assyrian capital *Ashur*—named the same as the land Assyria itself and its national god *Ashur*. (*Ressen,* which was also mentioned in Genesis and whose name meant 'Horse's Bridle', turned out to have been an Assyrian horse-raising site.)

The Assyrian discoveries offered not just corroboration of the Bible's historical veracity; the art and iconography also seemed to bear out other biblical aspects. Wall reliefs in Khorsabad and Nimrud depicted winged 'angels' (Fig. 21) akin to the divine attendants described in the vision of the Prophet Isaiah (6:2), or that of the Prophet Ezekiel's vision (1:5–8, where each had four wings but also four faces, one of which was an eagle's).

Figure 21

The discovered sculptures and wall depictions seemed to also support some of the statements attributed to Berossus regarding what one would describe today as 'bio-engineering gone awry'—of men with wings, bulls with human heads, and so on (as earlier quoted). In Nineveh and Nimrud, the entrances to the royal palaces were flanked by colossal stone sculptures of human-headed bulls and lions (Fig. 22); and on wall reliefs, there were images of divine beings dressed as fish (Fig. 23)—the very image of Oannes, exactly the way Berossus had described him.

Although it had been, when Berossus was writing, almost four centuries since Ashur, Nineveh, and other Assyrian centers had been captured and destroyed, and some three centuries since the same fate befell Babylon, their ruins were still visible without excavation—with the sculptures and wall reliefs for all to see, illustrating what Berossus was talking about. The ancient monuments literally corroborated what he had written.

* * *

Figure 22

Figure 23

But with all the unearthing of Assyria's and Babylon's grandeur, treasures, and larger-than-life art, the most important discoveries were the countless clay tablets, many assembled in actual libraries, where the first tablet on a shelf listed the titles of the other tablets on that shelf. Throughout Mesopotamia—indeed, throughout the ancient Near East—virtually each major urban center had a library as part of either the royal palace, the main temple, or both. By now, thousands upon thousands of clay tablets (or fragments thereof) have been found; most linger, untranslated, in museum and university basements.

Of the main libraries discovered, the one of the greatest consequence was Layard's find among the ruins of Nineveh: The great library of the Assyrian king Ashurbanipal (Fig. 24, from his monuments; 668–631 B.C.). It contained more than 25,000 (!) clay tablets. Their inscribed texts—all using cuneiform script—ranged from royal annals and records of workers' rations to commercial contracts and marriage and divorce documents, and included literary texts, historical tales, astronomical data, astrological forecasts, mathematical formulas, word lists, and geographic lists. And then there were rows of tablets with what

Figure 24

the archaeologists classified as 'mythological texts'—texts dealing with varied gods, their genealogies, powers, and deeds.

Ashurbanipal, it turned out, not only collected and brought back to Nineveh such historical and 'mythological' texts from every corner of his empire—he actually employed a legion of scribes to read, sort, preserve, copy, and translate into Akkadian the most important of them. (Depictions of Assyrian scribes show them dressed as dignitaries—attesting to their high status.)

Most of the tablets discovered at Nineveh were shared between the Ottoman authorities in Constantinople (Istanbul in today's Turkey) and the British Museum in London; some related tablets found their way to the main museums of France and Germany. In London, the British Museum engaged a young banknote engraver and amateur 'Assyriologist' named George Smith to help sort out cuneiform tablets. With a keen ability to recognize a particular characteristic of a cuneiform line, he was the first to realize that various fragmented tablets belonged together, forming continuous narratives (Fig. 25). There was one about a hero and a Flood, another about gods who created

Figure 25

Heaven and Earth and also Man. In a Letter to the Editor about it in a London daily, Smith was the first to draw attention to the similarities between the tales in those tablets and the biblical stories in Genesis.

Of the two ancient story lines, the one of the greatest religious ramifications was the one akin to the biblical tale of Creation; as it happened, the studies in that direction were led by a succession of scholars not in England but in Germany, where pioneering 'Assyriologists' such as Peter Jensen (*Kosmologie der Babylonier*), Hermann Gunkel (*Schöpfung und Chaos*), and Friedrich Delitzsch (*Das babylonische Weltschöpfungsepos*) utilized additional finds by the German archaeologists to form a more coherent text and understand its religious, philosophical, and historical scope.

At the British Museum in London, added to the tablets that Smith had been piecing together were new discoveries by a Layard trainee, Hurmuzd Rassam, at Nineveh and Nimrud. Pursuing the Creation story line, the Museum's Curator of Egyptian and Babylonian Antiquities, Leonard W. King, found that a veritable Epic of Creation was in fact inscribed on no less than seven tablets. His 1902 book, *The Seven Tablets of Creation,* concluded that a "Standard Text" had existed in Mesopotamia that, like Genesis, told a sequential tale of Creation—from Chaos to a Heaven and an Earth, and then on Earth from the Gathering of the Seas to the Creation of Man—not in the course of the biblical six days plus a day of self-gratification, but over six tablets plus a laudatory seventh.

The tale's ancient title, conforming to its opening words, was *Enuma elish* ("When in the Height Above"). Tablets from various sites seemed to have identical texts, except in the name by which the Creator Deity was called (the Assyrians called him **Ashur**, the Babylonians **Marduk**)—suggesting that they were all renditions adapted from a single canonical version in Akkadian. However, the occasional retention of some odd words, and names of celestial deities involved in the events—names such as **Tiamat** and **Nudimmud**—suggested that such an original version might not have been in Assyrian/Babylonian Akkadian, but in some other unknown language.

The search for origins, it was evident, was only beginning.

* * *

Back to Victorian England and George Smith: There and at that time, it was the other story line—the tale of the Deluge and a non-biblical 'Noah'—that captured popular imagination. Focusing his attention accordingly, the prolific George Smith, poring over thousands of tablet fragments from Nineveh and Nimrud and matching pieces together, announced that they belonged to a full length epic tale about a hero who discovered the secret of the Great Flood. The three cuneiform signs naming the hero were read by Smith *Iz-Du-Bar,* and Smith assumed that he was really the biblical ***Nimrod***—the "mighty hunter" who, per Genesis, had started the Assyrian kingdoms—in line with the name of the ancient site, Nimrud, where some of the tablets were found.

Smith's reading of the fragments, indicating the existence of an Assyrian Deluge story matching the one in the Bible, caused such excitement that the London newspaper *The Daily Telegraph* offered a grand prize of a thousand Guineas (a Guinea being worth more than a Pound Sterling) to anyone who would unearth missing fragments that would provide the full ancient story. Smith himself took up the challenge; he went to Iraq, searched the sites, and returned with 384 new fragmented tablets. They made possible the piecing together and sequencing of all twelve (!) tablets of the epic tale, including the crucial "Deluge Tablet," Tablet XI (Fig. 26). (As to the prize: It was the Museum that gratefully collected it, claiming that Smith went to Iraq while in its employ . . .)

One can only imagine the excitement of discovering the Hebrew Bible's tale of the Deluge and of Noah written down in other ancient languages unrelated to the Bible—a text of what has since been known as the Epic of Gilgamesh (the initial reading 'Izdubar' was in time dropped for the correct ***Gilgamesh***). But the euphoria was not without problems, among them the variety of gods involved in the event, compared to a sole Yahweh in the Bible.

Confounding the scholars, a king named Gilgamesh was nowhere listed as a Babylonian or Assyrian king. The hero Gilgamesh, scholars found, was identified in the very opening lines on Tablet I as king of ***Uruk,*** a city (according to the text) of wide walls and great ramparts.

Figure 26

But there was no ancient site by that name anywhere in Babylonia and Assyria. As the tale was pieced together, it was also realized that Gilgamesh himself was not the hero of the Flood. Being "two-thirds of him divine," his adventures were in search of immortality; and it was in the course of that search that he heard the tale of the Flood from one called ***Utnapishtim***—a Mesopotamian 'Noah' who had actually survived the catastrophe. So who was Gilgamesh—scholars and press wondered—if he was neither the biblical Noah nor the biblical/Assyrian Nimrod?

In 1876 Smith summed up his various findings in a short book, *The Chaldean Account of Genesis*. It was the first book to announce and compare the ancient texts discovered in Mesopotamia with the Creation and Deluge tales in the Bible. It was also Smith's last book: He died that same year, at the young age of 36; but it ought to be remembered that it was the ingenuity and findings of this self-taught master of Akkadian that served as the foundation for the subsequent myriad studies.

Those studies also uncovered the existence of yet another, more firsthand Deluge tale; its significance to our quest is that it had probably been a Berossus source. Titled in antiquity, as usual, after its opening words *Inuma ilu awilum* ("When the gods as men"), it has come to be known as the *Atra-H̲asis Epic,* after the name of its hero who tells the story of the Deluge firsthand—making him, ***Atra-H̲asis,*** the actual 'Noah' of this Deluge version. ***This is Noah himself speaking!***

For unclear reasons, it took time for scholarly attention to focus on this crucial text—crucial because in it Atra-H̲asis (= 'The Exceedingly Wise') tells *what had preceded the Deluge,* what brought it about, and what happened thereafter. In the course of piecing together the text's three tablets, a tablet-fragment marked "S" was essential for identifying the name *Atra-H̲asis;* the "S" stood for Smith; it was he who, before he died, had found the key to another amazing 'Babylonian' tale of gods, Man, and Deluge. As to the hero's name, it has been suggested, with little doubt, that *Atra-H̲asis,* transposed as *H̲asis-atra,* was the *Xisithros/Sisithros* in the Berossus Fragments—the tenth pre-Diluvial ruler in whose time the Deluge had occurred, just as Noah was the tenth biblical ancestor in the line of Adam!

(This name transposition is one of the reasons for linking Berossus to the *Atra-H̲asis* text. Another is the fact that it is only in this Mesopotamian version of the Deluge tale is there mention of the episode—mentioned by Berossus—of the townspeople questioning the building of the boat.)

It was all a wonder of wonders: Transcending time from the Babylonian Berossus in the 3rd century B.C. to the 19th century A.D., Bible-believing Western Man actually held in hand "*a Hebrew Deluge text written in cuneiform*" (as a Yale University publication called it in 1922, Fig. 27), inscribed on a tablet from a 7th century B.C. Assyrian library. This was an incredible time-bridging of at least 2,600 years; but that too proved to be just an interim way station on the march back in history.

* * *

YALE ORIENTAL SERIES · RESEARCHES · VOLUME V–3

A HEBREW DELUGE STORY
IN CUNEIFORM

AND OTHER EPIC FRAGMENTS IN
THE PIERPONT MORGAN LIBRARY

BY
ALBERT T. CLAY

NEW HAVEN
YALE UNIVERSITY PRESS
LONDON · HUMPHREY MILFORD · OXFORD UNIVERSITY PRESS
MDCCCCXXII

Figure 27

Once again, this Assyrian text appeared to have a similar or parallel Babylonian version. It too contained unfamiliar words and names, certainly not of Semitic-Akkadian provenance—gods named **Enlil**, **Enki** and **Ninurta**, goddesses named **Ninti** and **Nisaba**, divine groups called **Anunnaki** and **Igigi**, a sacred place named ***Ekur***. Where have they all come from?

The puzzlement was even greater when it became known that a partial *Atra-Hasis* tablet that had somehow made its way to the private Library of J. Pierpont Morgan in New York City circa 1897 contained a 'colophon'—a notation by the tablet's scribe—that dated the tablet to the *2nd millennium* B.C. Assyriologists were now looking at a leap back of 3,500 years!

Efforts to piece together as complete a text as possible from various tablets and several renditions resulted in tracing in the British Museum and in the Museum of the Ancient Orient in Istanbul, Turkey, of all *three* tablets (even though broken in parts) of that Babylonian version of *Atra-Hasis*. Fortunately, preserved in each one was the scribal statement giving his name, title, and date of completing the tablet (as this one at the end of the first tablet):

> Tablet 1. When the gods like men.
> Number of lines 416.
> [Copied] by Ku-Aya, junior scribe.
> Month Nisan, day 21,
> [of the] year when Ammi-Saduka, the king,
> made a statue of himself.

Tablets II and III were likewise signed by the same scribe and were also dated to a particular year in the reign of King Ammi-Saduka. It was not an unknown royal name: Ammi-Saduka belonged to the famed Hammurabi dynasty of Babylon; he reigned there from 1647 to 1625 B.C.

Thus, this Babylonian version of the Noah/Deluge tale was a thousand years older than Ashurbanipal's Assyrian version. And it too was a copy—of what original?

The incredulous scholars had the answers right in front of them. On one of his tablets Ashurbanipal boasted thus:

> The god of scribes has bestowed on me the gift
> of the knowledge of his art.
> I have been initiated into the secrets of writing.
> I can even read the intricate tablets in Shumerian.

> I understand the enigmatic words in the stone
> carvings from the days before the Flood.

Apart from disclosing the existence of a "god of scribes," here was a confirmation by an independent source, centuries before Berossus, of the occurrence of the Deluge, plus the detail that there had been "enigmatic words," *preserved in stone carvings "from the days before the Flood"*—a statement that matches and corroborates the Berossus assertion that the god Cronos "revealed to Sisithros that there would be a Deluge . . . and ordered him to conceal in Sippar, the city of the god Shamash, every available writing."

And then there is the prideful boast in the Ashurbanipal statement that he "could even read the intricate tablets in *Shumerian*."

Shumerian? The puzzled scholars—who had managed to decipher Babylonian, Assyrian, Old Persian, Sanskrit—wondered what Ashurbanipal was talking about. The answer, it was realized, had been provided by the Bible all along. Hitherto, the verses in Genesis 10:8–12 about the domains of the mighty hero Nimrod had inspired the decipherers of those ancient languages to name the mother tongue of Babylonian and Assyrian 'Akkadian', and served as a Discoverers' Map for the excavating archaeologists; now these verses also clarified the Shumerian mystery:

> He was first to be a hero in the Land;
> And the beginning of his kingdom:
> Babel and Erech and Akkad,
> all in the Land of ***Shine'ar***.

Sumer (or more correctly, ***Shumer***), was the biblical ***Shine'ar***—the very land whose settlers after the Deluge attempted to build a tower whose head could reach the heavens.

The search for Noah, it became clear, had to go to ***Shumer***—the biblical ***Shine'ar***—a land that undoubtedly predated the brought-back-to-light capitals of Babylon, Assyria, and Akkad. But which land, and where, had it been?

THE DELUGE

The common notion of the biblical Deluge (*Mabul* in Hebrew, from the Akkadian *Abubu*) is one of torrential rains whose outpour floods, overwhelms, and sweeps away everything on the ground below. In fact, the Bible (Genesis 7:11–12) states that the Deluge began when ***"all the water sources of the Great Deep burst apart."*** It was only after that (or as a result thereof) that "the sluices of the skies opened up, and the rain was upon the Earth forty days and forty nights." The Deluge ended in likewise sequence (Genesis 8:2–3), when first "the water sources of the Great Deep," and then "the sluices of the sky," shut down.

The varied Mesopotamian records of the Deluge describe it as an avalanche of rising waters *storming from the south,* overwhelming and submerging all as it rushed forth. The Akkadian version (Gilgamesh Tablet XI) states that the first manifestation of the Deluge was "a black cloud that arose from the horizon," followed by storms that "tore out the posts and collapsed the dikes." "For one day the South Storm blew, submerging the mountains, overtaking the people like a battle . . . seven days and [seven] nights blows the Flood-Wind as the South Storm sweeps the land . . . and the whole land was submerged like a pot."

In the Sumerian tale of the Deluge, howling winds are mentioned; rain is not: "All the windstorms, exceedingly powerful, attacked as one . . . For seven days and seven nights the Flood (**A.ma.ru**) swept over the land, and the large boat was tossed about by the windstorms on the great waters."

In *The12th Planet* and subsequent books, **I have suggsted that "the Great Deep," where the "South Storm" originated, was Antarctica;** and that the Deluge was *a huge tidal wave* caused by the slippage of the ice sheet off Antarctica—causing the abrupt end of the last Ice Age circa 13,000 years ago. (See also Fig. 43.)

IV

Sumer: Where Civilization Began

Sumer, it is now known, was the land of a talented and dexterous people in what is now southern Iraq. Usually depicted in artful statues and statuettes in a devotional stance (Fig. 28), it was the Sumerians who were the first ones to record and describe past events and tell the tales of their gods. It was there, in the fertile plain watered by the great Euphrates and Tigris Rivers, that Mankind's first known civilization blossomed out some 6,000 years ago—"suddenly," "unexpectedly," "with stunning abruptness," according to all scholars. It was a civilization to which we owe, to this day, virtually every 'First' of what we deem essential to an advanced civilization: The wheel and wheeled transportation; the brick that made (and still makes) possible high-rise buildings; furnaces and the kiln that are essential to industries from baking to metallurgy; astronomy and mathematics; cities and urban societies; kingship and laws; temples and priesthoods; timekeeping, a calendar, festivals; from beer to culinary recipes, from art to music and musical instruments; and, above all, writing and record keeping—it was all first there, in Sumer.

We now know all that thanks to the achievements of archaeology and the decipherment of ancient languages during the past century and a half. The long and arduous road by which ancient Sumer moved from complete obscurity to an awed appreciation of its grandeur has a

Figure 28

number of milestones bearing the names of scholars who had made the journey possible. Some, who toiled at the varied sites, will be mentioned by us. Others, who pieced together and classified fragmented artifacts during a century and a half of Mesopotamian archaeology, are too many to be listed.

And then there were the epigraphers—sometimes out in the field, most of the time poring over tablets in crammed museum or university quarters—whose persistence, devotion, and abilities converted pieces of clay incised with odd 'cuneates' into legible historical, cultural and literary treasures. Their work was crucial, for while the usual pattern of archaeological and ethnographic discovery has been to find a people's remains and then decipher their written records (if they had them), in the case of the Sumerians recognition of their language—even its decipherment—preceded the discovery of their land, *Sumer* (the common English spelling, rather than Shumer). And it was not because the language, 'Sumerian', preceded its people; on the contrary—it was because the language and its script lingered on after Sumer was long gone—just as Latin and its script had outlived the Roman empire thousands of years later.

The philological recognition of Sumerian began, as we have illustrated, not through the discovery of the Sumerians' own tablets, but through the varied use, in Akkadian texts, of 'loan words' that were not Akkadian; the naming of gods and cities by names that made no sense in Assyrian or Babylonian; and of course by actual statements (as that by Ashurbanipal) about the existence of earlier writings in 'Shumerian'. His statement was borne out by the discovery of tablets that rendered the same text in two languages, one Akkadian and the other in the mysterious language; then the next two lines were in Akkadian and in the other language, and so on (the scholarly term for such bilingual texts is 'interlinears').

It was in 1850 that Edward Hincks, a student of Rawlinson's Behistun decipherments, suggested in a scholarly essay that an Akkadian 'syllabary'—the collection of some 350 cuneiform signs each representing a full consonant + vowel syllable—must have evolved from a prior *non*-Akkadian set of syllabic signs. The idea (which was not readily accepted) was finally borne out when some of the clay tablets in the Akkadian-language libraries turned out to be *bilingual 'syllabarial' dictionaries*—lists that on one side of the tablet gave a cuneiform sign in the unknown language, and a matching list on the other side in Akkadian (with the signs' pronunciation and meaning added, Fig. 29). All at once, archaeology obtained *a dictionary of an unknown language!* In addition to tablets inscribed as a kind of dictionaries, the so-called Syllabaries, various other bi-lingual tablets served as invaluable tools in deciphering the Sumerian writing and language.

In 1869 Jules Oppert, addressing the French Society of Numismatics and Archaeology, pointed out that the royal title "King of Sumer and Akkad" found on some tablets provided the name of the people who had preceded the Akkadian-speaking Assyrians and Babylonians; they were, he suggested, the ***Sumerians***. The designation has been applied ever

Figure 29

since—although, to this day, museums and the media prefer to name their exhibits or title their articles and programs "Babylonian" or at best "Old Babylonian" rather than the unfamiliar "Sumerian." Though virtually everything that we consider essential to a developed civilization has been inherited from the Sumerians, many people still respond with a blank "Who?" when they hear the word 'Sumerian' . . .

The interest in Sumer and the Sumerians constituted a chronological as well as a geographical shift: From the 1st and 2nd millennia B.C. to the 3rd and 4th millennia B.C., and from northern and central Mesopotamia to its south. That ancient settlements lay buried there was indicated not only by the numerous mounds that were scattered over the flat mudlands, mounds that resulted from layers of habitats built upon layers (called strata) of the remains of previous habitats; more intriguing were odd artifacts that local tribesmen dug up out of the mounds, showing them to the occasional European visitors. What we know now is the result of almost 150 years of archaeological toil that brought to light, to varying degrees, Sumer's fourteen or so major ancient centers (map, Fig. 30), virtually all of which are mentioned in the ancient texts.

* * *

Systematic field archaeology of Sumer is deemed to have begun in 1877 by Ernest de Sarzec, who was then the French Vice-Consul in Basra, Iraq's southern port city on the Persian Gulf. (Rumors at the time were that having been fascinated by the local trade in finds, his real interest

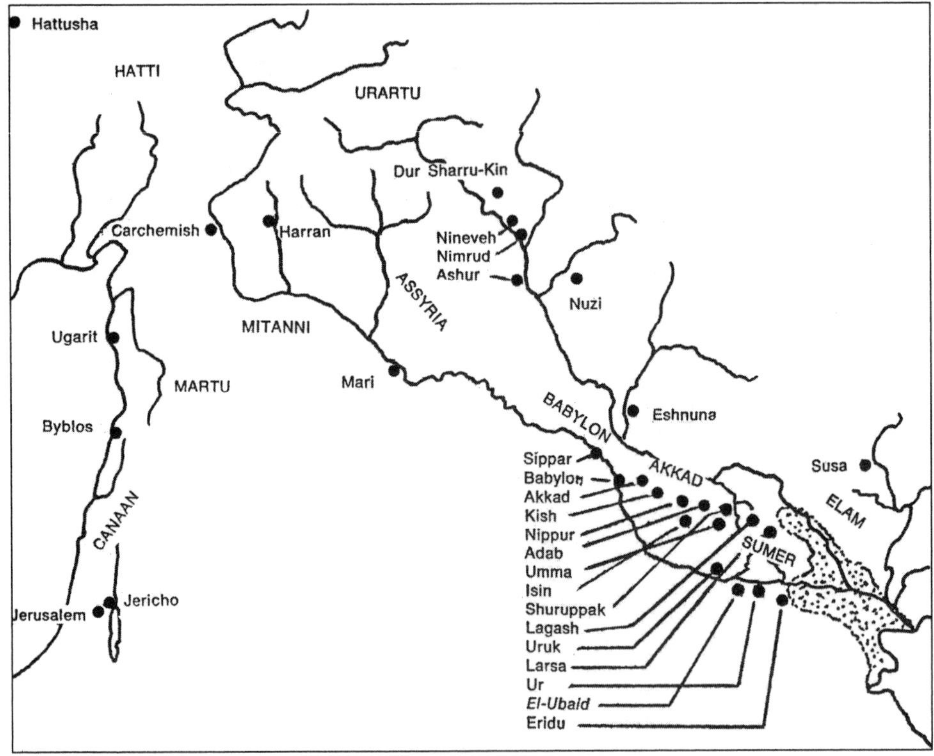

Figure 30

was in finding objects for private sale.) He started excavating at a site locally called *Tello* ('The Mound'). The finds there were so great—and they did go to the Louvre Museum in Paris, where they fill up galleries—and so inexhaustible, that French archaeological teams kept coming back year after year to this one site for *more than fifty years,* through 1933.

Tello turned out to be the sacred precinct, the *Girsu,* of a large Sumerian urban center called ***Lagash.*** Archaeological strata indicated that it had been continuously settled almost since 3800 B.C. Sculpted wall reliefs dating from a so-called Early Dynastic Period, stone sculptures bearing inscriptions in immaculate Sumerian cuneiform (Fig. 31), and a beautiful silver vase presented by a king named Entemena to his god (Fig. 32) attested the high level of Sumerian culture millennia ago. To top it all, more than 10,000 inscribed clay tablets were found in the city's library (the importance of which will be discussed later on).

Figure 31

Figure 32

Some inscriptions and texts named a continuous line of kings of Lagash who reigned from circa 2900 B.C. to 2250 B.C.—an uninterrupted reign of almost seven centuries. Clay tablets and commemorative stone plaques recorded large construction undertakings, irrigation and canal projects (and named the kings who initiated them); there was trade with distant lands, and even conflicts with nearby cities.

Most astounding were the statues and inscriptions of a king named Gudea (circa 2400 B.C., Fig. 33) in which he described the miraculous circumstances leading to the building of a complex temple for the god Ningirsu and the god's spouse, Bau. The task, detailed later on, involved divine instructions given in 'Twilight Zone' circumstances, astronomi-

Figure 33

cal alignments, elaborate architecture, the importation of rare building materials from distant lands, calendrial know-how, and precise rituals—all taking place some 4,300 years ago. The Lagash discoveries have been summed up by its last French excavator, Andrè Parrot, in his book *Tello* (1948).

A few miles northwest of the mounds of Lagash, a mound locally called Tell el-Madineh was located. The French excavators of Lagash peeked at it too; but there was not much to excavate, for the ancient city that had been there was, at some time, completely destroyed by fire. A few finds, however, helped identify that ancient city as ***Bad-Tibira***. The ancient city's Sumerian name, 'Bad Tibira', meant 'The Metalworking Fort'; as other discoveries clarified later, Bad-Tibira was indeed considered to have been a metalworking center.

A decade after de Sarzec began excavations at Lagash, a new major archaeological player joined the effort to uncover Sumer: The University of Pennsylvania in Philadelphia. It had been known, from preceding finds in Mesopotamia, that the most important religious center in Sumer was a city called ***Nippur;*** in 1887 John Peters, a professor of Hebrew at the university, succeeded in lining up academic support at the university and financial support from individual donors to organize an "archaeological expedition" to Iraq to find Nippur.

The location of Nippur seemed easy to guess: At the geographical center of southern Mesopotamia, a can't-be-missed huge mound rising some 65 feet above the mudplain was called *Niffar* by the locals; it fitted references to ancient Nippur as "Navel of the Earth." The University of Pennsylvania's Expedition conducted four excavation 'campaigns' at the site from 1888 to 1900, at first under the direction of John Peters, then under the leadership of Hermann Hilprecht, a German-born Assyriologist of international standing.

Nippur, the archaeologists ascertained, had been continuously settled from the *6th millennium B.C.* to about A.D. 800. The excavations focused at first on the city's Sacred Precinct whose location—as incredible as it may sound—was indicated on a millennia-old city map inscribed on a large clay tablet (Fig. 34, transcript and translation). There,

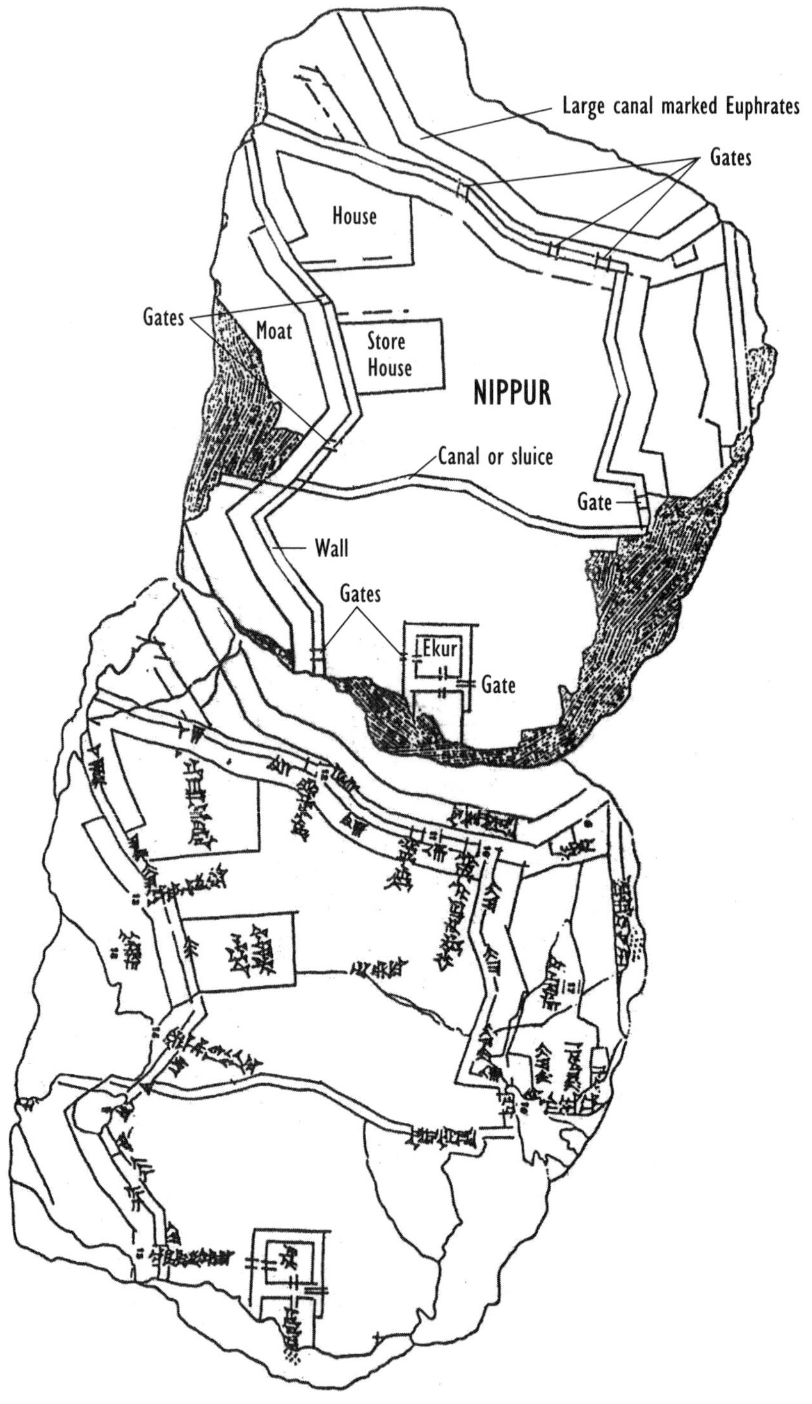
Large canal marked Euphrates
Gates
House
Gates
Moat
Store House
NIPPUR
Canal or sluice
Gate
Wall
Gates
Ekur
Gate

Figure 34

the remains of a high-rising ziggurat (step-pyramid) in the city's sacred precinct (reconstruction, Fig. 35) attested its dominance above the city. Called ***E.Kur*** (= 'House which is like a mountain'), it was the main temple dedicated to Sumer's leading god **En.lil** (= 'Lord of the Command') and his spouse **Nin.lil** (= 'Lady of the Command'). The temple, inscriptions stated, included an inner chamber in which "Tablets of Destinies" were kept. According to several texts, the chamber was the heart of the ***Dur.An.Ki*** (= "Bond Heaven-Earth')—a Command and Control Center of the god Enlil that connected Earth with the heavens.

The Expedition's finds at Nippur, deemed by some to be "of unparalleled importance," included the discovery of nearly 30,000 inscribed clay tablets (or fragments thereof) in a library of what had apparently been a special Scribal & Science quarter of the city, adjoining the Sacred Precinct. Hilprecht planned to publish no less than twenty volumes with the tablets' most important texts, many with "mythological" context,

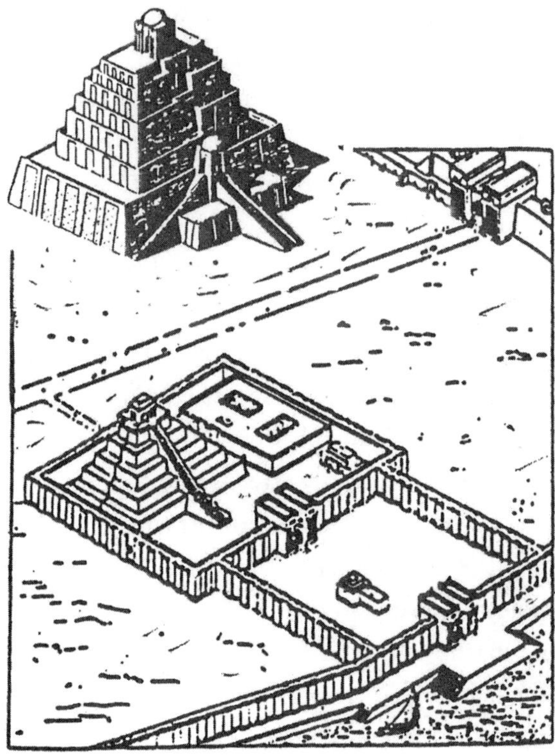

Figure 35

others dealing with mathematics and astronomy and dating back to the 3rd millennium B.C. Among the Nippur inscriptions that were transcribed, translated and published was a remnant of the ***original Sumerian tale of the Deluge,*** naming its "Noah" ***Ziusudra*** (= '[His] Lifedays Prolonged')—the equivalent of the Akkadian *Utnapishtim.*

In this Sumerian inscription (known to scholars by its reference number CBS 10673), it is the god Enki who reveals to his faithful follower Ziusudra a "secret of the gods"—that, at the instigation of an angry Enlil, the gods decided to "destroy the seed of Mankind by the Deluge" that was about to happen; and Enki ('Cronos' in the Berossus Fragments) instructs Ziusudra (the 'Xisithros' of Berossus) to build the salvaging boat.

But all the Expedition's plans were cut short by a spate of accusations by Peters that Hilprecht was providing misleading 'provenances' (discovery locations) for announced finds, and that Hilprecht had made a deal with the Turkish Sultan in Constantinople (today's Istanbul) to send most of the finds there—rather than to the university in Philadelphia—in exchange for the Sultan letting Hilprecht keep some finds as 'gifts' for his private collection. The controversy, which divided Philadelphia's highest echelons and made headlines in the *New York Times,* raged from 1907 to 1910. A commission of inquiry formed by the University in the end found the accusations of professional misconduct against Hilprecht to be "unsubstantiated"; but in fact many of the Nippur tablets did end up in Constantinople/Istanbul. Hilprecht's private collection ended up in Jena, Hilprecht's university town in Germany.

The University of Pennsylvania, through its Archaeological Museum, returned to Nippur only after World War II, in a joint expedition with Chicago University's Oriental Institute. The Peters-Hilprecht controversy is still regarded by historians as a major disruption of Near Eastern archaeology. But due to the ever-intervening Law of Unintended Consequences, in the end it led to one of the greatest advances in Sumerology, for it provided the first job to a young epigrapher named Samuel N. Kramer who then became an outstanding 'Sumerologist'.

* * *

The excavations at Lagash and Nippur, requiring continuous archaeological efforts year after year after year, revealed the existence of major urban centers in Sumer that rivaled in size the Babylonian and Assyrian sites in the north, even though the ones in Sumer were older by more than a thousand years. The existence of walled sacred precincts, each with a skyscraping ziggurat, indicated a high level of ancient building technology that preceded and served as a model for the Babylonians and Assyrians. The *ziggurats*—literally 'That which rises high'—rose in several steps (usually seven) to heights that could reach 90 meters. They were built of two kinds of mud bricks—sun-dried for high-rise cores, and kiln-burned for extra strength for stairways, exteriors, and overhangs; the size, shape, and curvature of the bricks varied to fit their function; and they were held together with bitumen as mortar. (Modern laboratory tests show that kiln-burnt mud bricks are fivefold stronger than sun-dried ones.)

The discovered ziggurats literally confirmed the biblical statement in Genesis 11:1–4 regarding the construction methods of the settlers in Shine'ar after the Deluge:

> And the whole Earth was of one language
> and one kind of words.
> And it came to pass,
> as they journeyed from the east,
> that they found a plain in the land of Shine'ar
> and they settled there.
> And they said unto each other:
> *Come, let us make bricks,*
> *and burn them thoroughly.*
> *And the brick served them for stone,*
> *and bitumen served them for mortar.*
> And they said:
> Come, let us build us a city,
> and a tower whose head shall reach heaven.

In lands like Canaan, where stones were used for building and lime is still used as mortar (for they lack bitumen), the reference to bricks and brick-making technology ("burn them thoroughly") and to bitumen (which seeps out of the ground in southern Mesopotamia)—represent a remarkably detailed and amazing knowledge of past events in a stoneless land like Sumer. Uncovering ancient Sumer, the archaeologists' spades were corroborating the Bible.

Beside the various technological accomplishments of those settlers in the plain between the Euphrates and Tigris Rivers—they also included the wheel and wagon, the kiln, metallurgy, medicines, textiles, multicolored apparel, musical instruments—there were countless other 'firsts' of what are still deemed essential aspects of an advanced civilization. They included a mathematical system called sexagesimal ('Base 60') that initiated the circle of 360°, timekeeping that divided day/night into 12 'double-hours', a luni-solar calendar of 12 months properly intercalated with a 13th leap month, geometry, measurement units of distance, weight and capacity, an advanced astronomy with planetary, star, constellation, and zodiacal knowledge, law codes and courts of law, irrigation systems, transportation networks and customs stations, dance and music (and musical notes), even taxes—as well as a social organization based on kingship and a religion centered at temples with prescribed festivals and a specialized priesthood. Additionally, the existence of scribal schools and temple and royal libraries indicated astounding levels of intellectual and literary achievements.

The Sumerologist Samuel Noah Kramer, in his trailblazing book *History Begins At Sumer* (1956), described twenty-seven of those Firsts, including the First Legal Precedent, the First Moral Ideals, the First Historian, the First Love Song, the First 'Job', and so on—all culled from Sumerian inscribed clay tablets. Actual archaeological finds of artifacts, and pictorial depictions, enhance and affirm that extensive textual record.

The realization in Europe and America of all of that served to increase the pace of uncovering Sumer; and the more archaeologists dug, the more they found themselves facing earlier and earlier times.

A site, called Bismaya, was excavated by an expedition of the University of Chicago. It was an ancient Sumerian city called ***Adab***.

Remains of temples and palaces were found there, with objects bearing votive inscriptions; some identified a king of Adab named Lugal-Dalu, who reigned there circa 2400 B.C.

At mounds grouped around the locally named Tell U<u>h</u>aimir, French archaeologists uncovered the ancient Sumerian city of ***Kish,*** with remains of two ziggurats; they were built of unusual convex bricks; a tablet inscribed in early Sumerian script identified the temple as dedicated to the god **Ninurta**, Enlil's warrior son. The earliest ruins, dated to the Very Early Dynastic period, included a palace of "monumental size"; the building was columned—a rarity in Sumer. The finds in Kish included remains of wheeled wagons and metal objects. Inscriptions identified two kings by their names—Mes-alim and Lugal-Mu; it was later determined that they reigned at the start of the 3rd millennium B.C.

Excavations at Kish were resumed after World War I by Chicago's Field Museum of Natural History and Oxford's Ashmolean Museum. Among their finds were some of the earliest examples of cylinder seal impressions. (In 2004 the Field Museum launched a project to unify, digitally on computers, *the more than 100,000 Kish artifacts* that have been dispersed between Chicago, London, and Baghdad.)

* * *

In the 1880s a site called Abu <u>H</u>abbah drew the attention of L. W. King of the British Museum when "interesting tablets"—dug up at the site by local plunderers—were offered for sale. A colleague, Theophilus Pinches, correctly identified the city as ancient ***Sippar***—the very city of the god Shamash, mentioned by Berossus in the story of the Flood!

The site was briefly excavated by Layard's assistant Hormuzd Rassam; one of the best known finds there has been a large stone tablet depicting none other than the god Shamash, sitting on his canopied throne (Fig. 36). The accompanying inscriptions identified the king being presented to the god as King Nabu-apla-iddin, who in the 9th century B.C. refurbished the Shamash temple in Sippar.

The city's twin mounds were more thoroughly excavated in the 1890s by a joint expedition of the Deutsche Orient Gesellschaft and the

Figure 36

Ottoman Antiquities Service. They not only discovered undisturbed hoards of textual tablets—shared between Berlin and Constantinople—but also some of the tablets' oldest and oddest libraries: The tablets were kept in 'pigeonhole' compartments cut into the mud-brick walls, rather than (as in later periods) on shelves. The library's texts included tablets whose colophons explicitly stated that those were copies of texts from earlier tablets coming from Nippur, from a city called Agade, and from Babylon—or found in Sippar itself; *among them were tablets belonging to the* ***Sumerian*** *Atra-H̲asis text!*

Did that indicate that Sippar had been an early repository of "writings," as the statements by Berossus have suggested? No certain answers can be given, except to quote Berossus again: First, 'Cronos' ordered Xisithros "to *dig a hole* and to bury all the writings about the Beginnings, Middles, and Ends, in Sippar, the city of the Sun god [Shamash]." Then, the Flood's survivors "came back to Babylon, they dug up the writings from Sippar, founded many cities, set up shrines, and once again established Babylon." Was the unique storage in cutout compartments a reminder of the "digging of holes" to preserve the most ancient tablets? We can only wonder.

At Sippar, the tale of the Deluge began to assume physical reality; but it was only the beginning.

In the decade preceding World War I, German archaeologists, under the auspices of the Deutsche Orient Gesellschaft, began excavating at a site locally named Fara. It was an important Sumerian city called ***Shuruppak,*** which had been settled well before 3000 B.C. Among its interesting features were buildings that were, without doubt, public facilities, some serving as schools with built-in mud-brick benches. There were plenty of inscribed tablets whose contents threw light on daily life, the administration of laws, and the private ownership of houses and fields—tablets that mirrored urban life five thousand years ago. Inscribed tablets asserted that this Sumerian city had a pre-Diluvial predecessor—a place that played a key role in the events of the Deluge.

The discoveries there stood out by their unusual hoard of *cylinder seals* or their impressions—a unique Sumerian invention that, as the cuneiform script, was in time adopted throughout the ancient lands. These were cylinders (mostly an inch or two in length) that were cut from a stone (often semiprecious), into which the artisan engraved a drawing, with or without accompanying writing (Fig. 37). The trick was

Figure 37

to engrave it all in reverse, as a negative, so that when it was rolled on wet clay the image was impressed as a positive—an early 'rotary press' invention. These cylindrical works of art are called 'seals' because that was their purpose: The seal's owner impressed it on a lump of wet clay that sealed a container of oil or wine, or on a clay envelope to seal a clay letter inside. Some seal impressions had already been found in Lagash, bearing the name of their owner; but the ones in Fara/Shuruppak exceeded 1,300 in number, and in some cases were from the earliest times.

But no less an amazing aspect of uncovering Shuruppak was its very finding—for, according to Tablet XI of the Akkadian version of the Epic of Gilgamesh, ***Shuruppak was the hometown of Utnapishtim, the 'Noah' of the Deluge!*** It was there that the god Enki revealed to Utnapishtim the secret of the coming Deluge and instructed him to build the salvage boat:

> Man of Shuruppak, son of Ubar-Tutu:
> Tear down the house, build a ship!
> Give up possessions, seek thou life!
> Forswear belongings, keep soul alive!
> Aboard ship take thou the seed of all living things.
> That ship thou shalt build—
> Its dimensions shall be to measure.

(Enki, it will be recalled, was reported to have been the revealer of the gods' secret decision also in the Sumerian text mentioned earlier.)

The discoveries of and at Shuruppak, together with those at Sippar, transformed the Deluge tale from legend and 'myth' to a physical reality. In *Divine Encounters* I have concluded, based on ancient data and modern scientific discoveries, that the Deluge was a colossal tidal wave caused by the slippage of the eastern Antarctic ice sheet off that continent.

World War I (1914–1918) interrupted those and other archaeological explorations in the Near East, which was part of the Ottoman empire until its dismemberment after the war. Mesopotamia was left in the hands of local excavators—both official, and (mostly) private site-robbers. Some of the finds did reach the Museum of the Ancient Orient in Constantinople/Istanbul, revealing that during the war years excavations in Iraq had taken place at Abu Habbah, ancient Sippar; but there was so much to uncover there, that varied excavations have continued into the 1970s—almost a full century after excavations there began.

* * *

A continuous and most determined series of excavations, lasting from the end of World War I until the outbreak of World War II in 1939 (and resumed in 1954) took place at a southern Sumerian site locally called Warka—*the very **Uruk** of the Epic of Gilgamesh, the **Erech** of the Bible!*

Adopting an excavating technique that cut a vertical shaft through all the strata, the German archaeologists of the Deutsche Orient Gesellschaft were able to see at a glance the site's settlement and cultural history—from the latest settlement at the top to a beginning in the *4th millennium B.C.* at the bottom. At all times since at least 3800 B.C., it appeared, every power from Sumerian, Akkadian, Babylonian, and Assyrian to Persian, Greek, and Seleucid wanted to leave a footprint at Uruk. Uruk, it was apparent, was a special place.

At Uruk the German archaeologists found several 'firsts'—the first items of colored pottery baked in a kiln, the first use of a potter's wheel, the first objects of metal alloys, the first cylinder seals, and the first inscriptions in the pictorial predecessor of cuneiform. Another first was a pavement made of limestone blocks, part of an unusual use of stones rather than mud bricks for construction—unusual because the stones had to be brought from mountains situated more than fifty miles to the east. The archaeologists described some of the city's stone buildings as of "monumental proportions."

A massive wall surrounded the city—the archaeologists found its remains over a length of more than 10 kilometers (more than six miles). It embraced the city's two sections—a residential one, and a sacred precinct where they discovered the earliest 'ziggurat'—a platform, raised in stages serving as a base for a temple. By the time of its excavation it was more like an artificial mound of no less than seven strata of rebuilding. On top, upon an artificially made platform, there stood a temple. Called *E.Anna* (= 'House/Abode of Anu') it is also known to archaeologists as the White Temple because—another unusual feature, a first—it was painted white (Fig. 38, a reconstruction). Next to the E.Anna were remains of two other temples. One, painted red, was dedicated to the goddess **In.anna**, 'Anu's Beloved' (better known by her later Akkadian name *Ishtar*). The other standing was a temple dedicated to the goddess Ninharsag.

Without doubt, ***the archaeologists' spade brought to light the city of Gilgamesh,*** who had reigned there circa 2750 B.C. (or even earlier by another chronology). The archaeologists' finds echoed literally the very words of the Epic of Gilgamesh—

> About all his toil he [Gilgamesh]
> engraved on a stone column:
> Of ramparted Uruk, of the wall he built,
> Of hallowed E.Anna, the pure sanctuary.
> Behold its outer wall, which is like a copper band,
> Peer at its inner wall, which none can equal!
> Gaze upon the stone platform, which is of old;
> Go up and walk around on the walls of Uruk,
> Approach the E.Anna and the dwelling of Ishtar!

Among the "small finds" in the 3200–2900 B.C. stratum were sculpted objects that were designated 'The Most Prized' in the Iraq

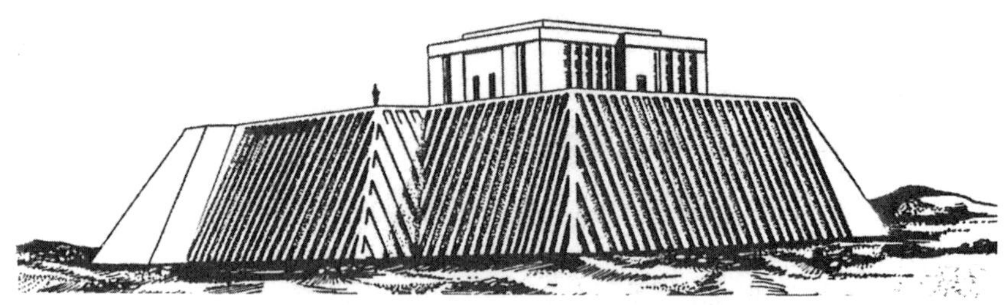

Figure 38

Museum in Baghdad—a life-size marble sculpture of a woman's head (Fig. 39)—nicknamed "The Lady from Uruk"—that had once been fitted with a golden headdress and eyes made of precious stones, and a large (more than 3 ft. high) sculpted alabaster vase that depicted a procession of adorants bearing gifts to a goddess. All at once, Sumer's art of more than 5,000 years ago matched the beauty of Greek sculpture of 2,500 years later!

At the southernmost part of Sumer, where the Tigris and Euphrates Rivers come together in marshlands bordering the Persian Gulf, a site locally called Abu Shaḫrain had attracted the attention of the British Museum as early as 1854. One of its experts, J. E. Taylor, reported after preliminary diggings that the effort was "unproductive of any very important results." He did bring back with him some of the "unimportant" finds—some mud bricks with writing on them. Fifty years later, two French Assyriologists determined from those bricks that the site was ancient ***Eridu;*** its name meant 'House in the Faraway Built', and ***it was Sumer's first city.***

Figure 39

It took two world wars and the time in-between for the first methodical and continuous archaeological excavations to take place at the site, under the auspices of the Iraqi Directorate General of Antiquities. As the archaeologists dug away occupation stratum after occupation stratum from the latest on top to the earliest at the bottom, they uncovered no less than seventeen levels above the first one; they could count time backward as they kept excavating: 2500 B.C., 2800 B.C., 3000 B.C., 3500 B.C. When the spades reached the foundations of Eridu's first temple, the date was circa 4000 B.C. Below that, the archaeologists struck virgin mud-soil.

The city's original temple, which had been rebuilt time and time again, was constructed of fired mud bricks and rose upon an artificial level platform. Its central hall was rectangular in shape, flanked on its two longer sides by a series of smaller rooms—a model of other temples in millennia to come. At one end there was a pedestal, perhaps for a statue. At the other end a podium created an elevated area; the astounded excavators discovered there, at levels VI and VII, large quantities of fish bones mixed with ashes—leading to the suggestion that fish were offered there to the god.

The excavators should not have been puzzled: The temple was dedicated to the Sumerian god **E.A**, whose name meant "He Whose Home Is Waters." It was he, as his autobiography and many other texts make clear, who had waded ashore from the Persian Gulf at the head of fifty *Anunnaki* spacemen who had come to Earth from their planet. Customarily depicted with outpouring streams of water (Fig. 40), ***it was he who was the legendary Oannes.*** In time—as explained in the preamble of the *Atra-Hasis* epic—Ea was granted the epithet **En.ki**—'Lord [of] Earth'. And it was he who had alerted Utnapishtim/Ziusudra of the coming Deluge, instructing him to build the waterproof boat and be saved.

* * *

Though wholly unintended, the unearthing of Eridu opened the way to archaeological confirmation of one of Sumer's most basic 'myths'—the

Figure 40

coming of the Anunnaki to Earth and the establishment by them of ***Cities of the Gods in pre-Diluvial times.***

It was in 1914 that one of the early 'Sumerologists', Arno Poebel, made known the astounding contents of a tablet kept in a fragments-box catalogued 'CBS 10673' in the collection of the Philadelphia University Museum. Less than half preserved (Fig. 41), this remainder of *the original Sumerian Deluge record* provides on the obverse side the bottom part of the first three columns of text; and turned over, it retains on the reverse the upper part of columns IV–VI.

The extant lines in the latter section relate how Ziusudra had been forewarned (by the god Enki) about the Deluge and the boat he was instructed to build, how the Deluge had raged for seven days and seven nights, and how the gods led by Enlil granted Ziusudra "life, like a god"—thus his name, "He of Prolonged Lifedays."

The obverse columns I–III, however, considerably expand the tale. The text describes the circumstances of the Deluge and the events that preceded it. Indeed, the text harks back to the time when the Anunnaki had come to Earth and settled in the ***Edin***—a tale that has led some to call this text *The Eridu Genesis.* It was in those early days, when the

Figure 41

Anunnaki brought 'Kingship' down from Heaven, the text asserts (in column II) that five Cities of the Gods were founded:

> After the [. . .] of Kingship
> was brought down from heaven,
> After the lofty crown and throne of kingship
> were lowered from heaven,

[. . .] perfected the [. . .],
[. . .] founded [. . .] cities in [. . .],
Gave them their names,
allocated their pure places:

The first of these cities, Eridu,
to the leader, Nudimmud, was given.
The second, Bad-Tibira, he gave to Nugig.
The third, Larak, to Pabilsag was given.
The fourth, Sippar, he gave to the hero Utu,
The fifth, Shuruppak, to Sud was given.

The disclosure that some time after they had arrived on Earth—but long before the Deluge—the Anunnaki established five settlements is a major revelation; that the cities' names, and names of their god-rulers, are stated, is quite astounding; but what is even more amazing about this list of Cities of the Gods is *that four of their sites have been found and excavated by modern archaeologists!* With the exception of ***Larak,*** whose remains have not been identified though its approximate location has been ascertained, ***Eridu, Bad-Tibira, Sippar,*** and ***Shuruppak*** have been found. Thus, as Sumer, its cities, and its civilization have been brought back to light, not only the Deluge but events and places from before the Deluge emerged as historical reality.

Since the Mesopotamian texts assert that the Deluge devastated the Earth and all upon it, one may well ask how those cities were still extant after the Deluge. For the answer—provided by the same Mesopotamian texts—we have to pull away the curtains of time and obscurity and reveal the full story of the *Anunnaki,* "Those Who From Heaven to Earth Came."

As before, it will be the ancient texts themselves that will tell the story.

THE LAND OF 'EDEN'

The name ***Shumer*** by which southern Mesopotamia was known in ancient times stems from Akkadian inscriptions about the kingdom of 'Shumer and Akkad'—a geopolitical entity formed after the installation of the Semitic-speaking Sargon I (*Sharru-kin* = 'The Righteous King') as ruler of Greater Sumer, circa 2370 B.C. (When the kingdom of David split up after his death to the kingdoms of Judea and Israel, the northern region was affectionately called *Shomron*—'Little Shumer'.)

Stemming from the Akkadian (and Hebrew) verb meaning 'to watch/to guard', the name *Shumer* identified the realm as "Land of the Watchers" or "Land of the Guardians"—the gods who watch over and safeguard Mankind. The term matched the ancient Egyptian word for 'gods'—***Neteru***—which stemmed from the verb NTR and meant "to guard, to watch over." According to Egyptian lore, the Neteru came to Egypt from ***Ur-Ta,*** the 'Ancient Place'; their hieroglyphic symbol was a miners' ax:

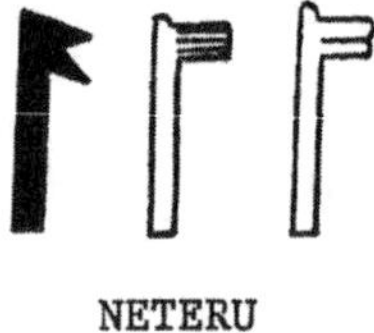

NETERU

Before Sumer & Akkad, when there were only Cities of the Gods in the land, it was called ***E.din***—'Home/Abode of the Righteous Ones"—the biblical *Eden;* the term stemmed from the determinative **Din.gir** that preceded gods' names in Sumerian. Meaning the 'Just/Righteous Ones', its pictographic depiction displayed their two-stage rocketships:

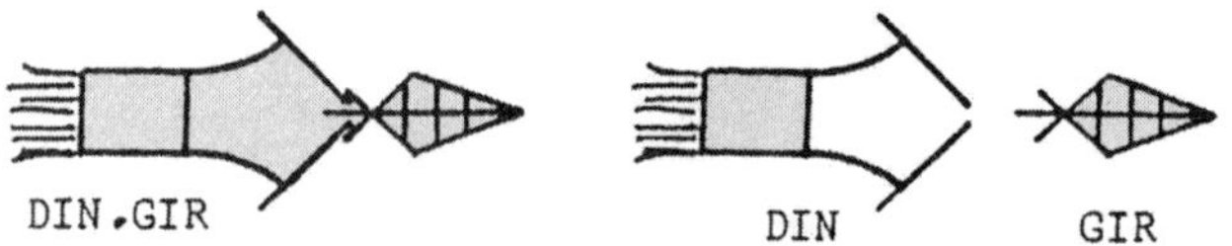

V

When Kingship Was Brought Down from Heaven

Cities—urban centers of population—are a hallmark of advanced civilization. The Sumerian tablet that relates the tale of the first five cities on Earth is thus the record of the start of advanced civilization on Earth.

Cities imply specialization between farming and industry, have buildings and streets and marketplaces, develop commerce and trade, entail transportation and communications, need record-keeping—reading, writing, and arithmetic. They also require an organized society and laws, have an administrative hierarchy, appoint or anoint a Chief Executive; in Sumer, and thereafter virtually everywhere else, that was a **Lu.gal**—literally "Big Man," rendered 'King' in translation. The Sumerians denoted these elements of advanced knowledge and the sum total of civilization in the term **Nam.lugal.la**, a term usually translated 'Kingship'. ***And Kingship, the Sumerians asserted, was brought down to Earth from the heavens.***

Held to be a divine institution, Kingship required that the king, to be legitimate, had to be chosen (or actually anointed) by the gods. Accordingly, throughout the ancient world, the succession of kings was meticulously recorded in King Lists. In Egypt, as we have seen, they were arranged by dynasties; in Babylonia and Assyria, in Elam and Hatti and Persia and beyond—and in the Bible with its two Books of Kings—the King Lists named successive rulers, giving their lengths of reign and occasionally a brief biographical note. In Sumer, where kings ruled in numerous city-states, the main list was arranged by the royal cities that served as the land's central or 'national' capital at any given time—a function that rotated from one major city to another. And Sumer's most famous and best preserved King List begins with the statement "When Kingship was brought down from Heaven"—a statement that matches the opening verses of the tale of the pre-Diluvial Cities of the Gods, that we have just quoted: "After the [. . .] of Kingship was brought down from heaven, after the lofty crown and throne of Kingship were lowered from heaven."

Those assertions, it should be clear, were not meant just to enshrine Kingship with divine status; a fundamental tenet of Sumerian history and teachings was that Kingship was *actually,* and not just figuratively, brought down to Earth from the heavens—that the **Anunnaki** (= 'Those who from Heaven to Earth came') actually began their civilized presence on Earth in five settlements, as stated in tablet CBS-10673. Though the name of the god who made the grants is missing in that tablet, one can say with certainty that it was Enlil, who followed Enki in coming to Earth—a detail recognized by the statement that "The first of these cities, Eridu, to *the leader* Nudimmud (= Ea/Enki) was given." Furthermore, each one of the others who were granted a city—Nugig (= the Moon god Nannar/Sin), Pabilsag (= Ninurta), Utu (= Shamash), and Sud (= 'The Medic', Ninmah)—was not just a high-ranking member of the Sumerian pantheon, but was related to Enlil. It was after Enlil's arrival that Enki's initial outpost (Eridu) was expanded to five (then eight) full-fledged settlements.

The connection between those first of Cities of the Gods and the bringing down of civilization to Earth from the heavens is restated in several other Sumerian documents dealing with pre-Diluvial events. Two of the key documents can be seen by anyone who visits the Ashmolean Museum of Art and Archaeology in Oxford, England—a museum that traces its beginning to the donation, in 1683, by Elias Ashmole, of twelve cartloads of antiquarian collectibles that the official history of the Museum describes as "a *Noah's ark* of rarities." The original collection diversified and grew over the centuries to become an official institution of the University of Oxford. Throngs do not wait in line to enter it; it has no *Mona Lisa* to attract multitudes or moviemakers. But among the objects it houses are two priceless archaeological finds of utmost importance to the history of Mankind and our planet; and both record the Deluge, alias "*Noah's* Flood." ***They, or copies of them, in all probability served as a source for the writings of Berossus.***

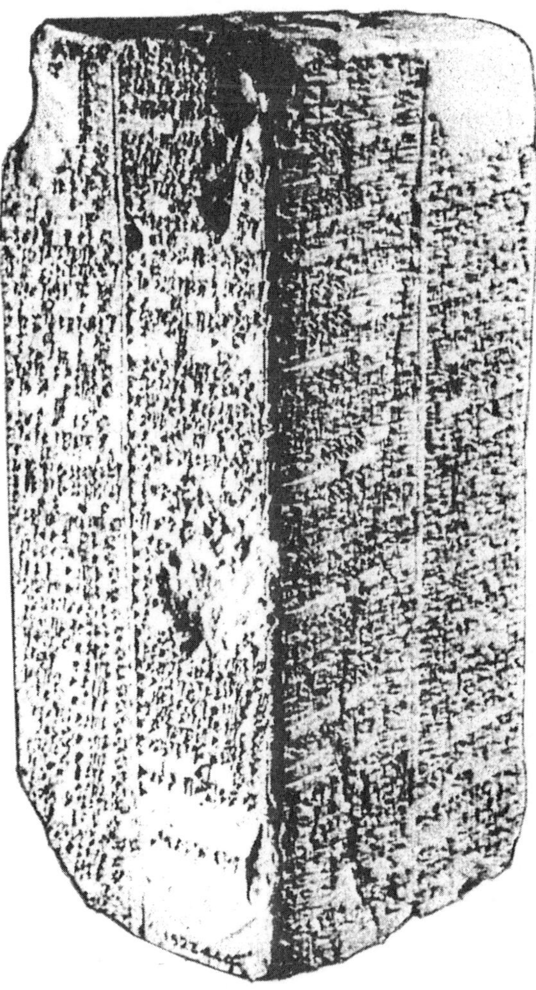

Figure 42

Catalogued as WB-62 and WB-444 by Stephen Langdon in *Oxford Editions of Cuneiform Texts* (1923), the two clay Sumerian artifacts belong to the private collection that Herbert Weld-Blundell—an English journalist, explorer, and archaeologist—donated to the museum in 1923. WB-444 is the better known of the two, for while WB-62 looks like the 'usual' kind of clay tablet, WB-444 is a rare, remarkable and beautiful four-sided prism of baked clay (Fig. 42). Known as ***The Sumerian King List,*** it details how Sumer's capital was first in the city of Kish, then moved to Uruk and to Ur, changed to Awan, returned to Kish, transferred to H̲amazi, returned to Uruk and then to Ur, and so on, ending in the city called Isin. (The last entry dates the document to a king named Utu-Hegal, who reigned in Uruk circa 2120 B.C.—more than 4,100 years ago.)

But those kings, the prism's text clearly states, began to reign only *after* the Deluge, "when Kingship was lowered [again] from heaven." The initial portion of the prism lists kings in five *pre*-Diluvial Cities of the Gods, assigning to each ruler lengths of reign that baffle scholars. This is what it says:

> **Nam.lugal an.ta e.de.a.ba**
> [When] Kingship from heaven was lowered,
> **Erida.ki nam.lugal.la**
> In Eridu was Kingship.
> **Erida.ki A.lu.lim Lugal**
> [In] Eridu Alulim was king,
> **Mu 28,800 i.a**
> 28,800 years [he] reigned.
> **A.lal.gar mu 36,000 i.a**
> Alalgar 36,000 years reigned;
> **2 Lugal**
> 2 kings
> **Mu.bi 64,800 ib.a**
> Its 64,800 years reigned.

And, continued in translation only:

Eridu was dropped,
Kingship to Bad-Tibira was carried.
In Bad-Tibira Enme.enlu.anna reigned 43,200 years;
Enme.engal.anna reigned 28,800 years;
Dumuzi, a shepherd, reigned 36,000 years;
3 kings reigned its 108,000 years.

Bad-Tibira was dropped,
Kingship to Larak was carried.
In Larak En.sipazi.anna reigned 28,800 years;
1 king reigned its 28,800 years.

Larak was dropped,
Kingship to Sippar was carried.
In Sippar Enme.endur.anna became king,
and reigned for 21,000 years.
1 king reigned its 21,000 years.

Sippar was dropped,
its Kingship to Shuruppak was carried.
In Shuruppak Ubar-Tutu became king,
and reigned its 18,600 years.

5 cities were they;
8 kings reigned 241,200 years.
The Flood swept thereover.

After the Flood had swept thereover,
When Kingship was lowered [again] from heaven,
the Kingship was in Kish.
etc., etc.

This usual rendering of WB-444's first lines is misleading in one key respect: In the original clay document, the numbers of the lengths of reign are given in *Sar* units (using the numeral sign for '3,600'): Alulim's reign in Eridu is not stated '28,800 years', but "8 Sars"; Alalgar's reign was not '36,000 years', but "10 Sars," and so on down the list of the pre-Diluvial rulers. ***The Sar units in this prism are the very Saros of Berossus.*** Significantly, the *Sar* unit of reign applied only to the pre-Diluvial rulers in the Cities of the Gods; the unit of count changes to regular numbers in the post-Diluvial part of the document.

No less significant is the fact that this list of rulers *names the very same first five cities, in the exact same order, as does tablet CBS-10673;* but rather than naming the gods whose 'cult center' each city was, WB-444 lists the 'kings'—administrators—of each such city. As a major study by William W. Hallo (*The Antedeluvial Cities*) has established, both documents *record a canonical tradition regarding the start of civilization* ('Kingship') on Earth, beginning in Eridu and ending in Shuruppak at Deluge time.

One cannot fail to note that WB-444 does not mention the hero of the Deluge, Ziusudra, among the eight kings it names. Embracing the cities and reigns from the start in Eridu to the deluvial finale in Shuruppak, its list ends with Ubar-Tutu and not with Ziusudra; but as tablet XI of the Epic of Gilgamesh clearly stated, the hero of the Deluge, Utnapishtim/Ziusudra, was the last ruler of Shuruppak, and he was the son and successor of Ubar-Tutu.

Various discoveries of other complete or fragmented similar tablets (such as UCBC 9-1819, Ni-8195, Baghdad 63095) leave no doubt that a canonized text, from which copies and copies of copies were made, did exist in regard to the pre-Diluvial Cities of the Gods and their rulers; and in the course of such copying, errors and omissions took place. One such little-known tablet is kept in a private collection in the Karpeles Manuscript Library Museum in Santa Barbara, California. It too names 8 kings in 5 cities, but its different reign lengths add up to "10 great Sars + 1 Sar + 600 x 5," which equals only 222,600 years.

The glaring omissions of Ziusudra are corrected in another tablet (British Museum K-11624). Called by some scholars *The Dynastic Chronicle,* it lists 9 kings in the first five cities, again with somewhat different *Sar* numbers—Alulim 10 (= 36,000), Alalgar 3 (= 10,800) instead of 28,800, and so on—but correctly ending with two kings in Shuruppak: Ubartutu with 8 *Sars* (= 28,800 years) and Ziusudra with 18

Sars (= 64,800 years). This tablet adds after the total of "5 cities, 9 kings, 98 *Sars*" (= 352,800 years) a brief explanations for the Deluge: "Enlil took a dislike to mankind; the clamor they made kept him sleepless" . . .

The tablet that gives the most accurate list of ten rulers, matching the Berossus list, is the Ashmolean Museum's tablet WB-62; its *Sar* units for the pre-Diluvial list parallel the *Saros* units of Berossus, though with different individual reign periods. It differs from WB-444 in listing not five but six cities, adding the city of ***Larsa*** (and with it two rulers) to the pre-Diluvial list—resulting in the full ten rulers, and ending correctly with Ziusudra at Deluge time. A comparison of WB-62 with the Greek fragments of Berossus (converting *Sars/Saros* to numbers of years) points strongly to this version as his principal source:

WB-62		Berossus	
Alulim	67,200	Aloros	36,000
Alalgar	72,000	Alaparos	10,800
[En]kidunu	72,000	Amelon	46,800
[. . .]alimma	21,600	Ammenon	43,200
Dumuzi	28,800	Megalaros	64,800
Enmenluanna	21,600	Daonos	36,000
Ensipzianna	36,000	Euedorachos	64,800
Enmeduranna	72,000	Amempsinos	36,000
Sukurlam (?)	28,800	Ardates (or Obartes)	28.800
Ziusudra	36,000	Xisuthros	64,800
Ten rulers	456,000	Ten kings	120 Shars = 432,000

Which of the various tablets that we have reviewed is the most accurate? The one that ends in Shuruppak and includes Ziusudra and his father/predecessor is plausibly the most reliable; with them, the list has ten pre-Diluvial rulers in six Cities of the Gods. The Bible too lists ten pre-Diluvial Patriarchs; though all were descendants of Adam through his grandson ***Enosh*** (Hebrew for 'Human') and not considered gods, the fact that they numbered ten and that the hero of the Deluge, Noah, was—like Ziusudra—the tenth, adds support to the Ten Rulers count as the correct one.

Despite the varying individual reign lengths, the various tablets unanimously agree that those successive rulers reigned from when "kingship was brought down from heaven" until "the Deluge swept thereover." Assuming that Berossus had reported the most reliable version, we also end up with his total of 120 *Sars* (= 432,000 years) as the correct combined total of the pre-Diluvial reigns—the time that had passed from when "kingship was brought down from heaven" until the Deluge. Thus, *if we could determine when the Deluge had occurred, we would obtain the date when the Anunnaki had arrived on Earth.*

That the number 120 appears in the biblical preamble to the tale of the Deluge (Genesis 6:3) might thus be more than a coincidence. The usual explanation, that it represents a limit on human longevity set by God at the time of the Deluge, is a dubious explanation in view of the fact that the Bible itself reports right thereafter that Shem, the eldest son of Noah, lived after the Deluge to the age of 600 years, his son Arpakhshad to 438, then Shelach to 433 years, and so on in descending longevities to 205 years for Terah, the father of Abraham; and Abraham himself lived to age 175. Moreover, a careful reading of the Hebrew shows that Genesis 6:3 states "and ***his*** years ***were*** one hundred and twenty." "Were" (not "will be"); and "his" can be understood as referring to the deity's count (in *Sar* years!) of his own presence on Earth from Arrival to Deluge. In earthly years, that would be 432,000 (120 x 3,600)—a statement matching the ten-kings/120-Sars of Berossus and the Sumerian King List.

Such a recollection of a pre-Diluvial 'Era of the gods' can explain the fact that the number 432,000 has been associated with divine duration in varied cultures, well beyond the boundaries of Mesopotamia. It forms, for example, the core of Hindu traditions about the Ages ('Yugas') of Earth, Mankind, and the gods: A Caturyuga ('Great Yuga') of 4,320,000 years was divided into four Yugas whose diminishing lengths were expressions of 432,000 years—the Golden 'Fourfold

Age' (432,000 x 4), the Threefold Age of Knowledge (432,000 x 3), the Twofold Age of Sacrifice (432,000 x 2), and finally our present era, the Age of Discord (432,000 x 1). According to the Egyptian priest Manetho, the "duration of the world" was 2,160,000 years; that equals five eras of 432,000, or 500 *Sar* years (3,600 x 500 = 2,160,000).

The 'Day of the Lord Brahma' of 4,320,000,000 years equaled 1,000 Great Yugas—reminding one of the biblical statement (Psalms 90:4) that in the eyes of God a thousand years are as just one day. In *Hamlet's Mill* (1977), Professors Giorgio de Santillana and Hertha von Dechend cite additional instances of 432,000 serving as "the point where myth and science join."

* * *

Modern scientific discoveries that have been presented in detail in *Genesis Revisited* and *Divine Encounters* have led me to conclude that the great Flood was a huge tidal wave caused by the slippage of the ice sheet off Antarctica. The elimination of that 'ice box', I have suggested, caused the abrupt end of the last Ice Age circa 13,000 years ago.

(The continent of Antarctica was discovered only in A.D. 1820; yet it was already shown on the A.D. 1513 map of the Turkish admiral Piri Re'is. As described in *Divine Encounters,* the slippage also explains the puzzle of other pre-discovery *mapas mundi,* such as the 1531 Orontius Finaeus map [Fig. 43] that shows Antarctica [box on the right] as though seen from the air and ***ice free*** [box on the left]; the contours of the Antarctican continent under the ice cover were determined by radar and other modern means only in 1958.)

The abrupt end of the last Ice Age has been the subject of numerous studies, including an intensive investigation during the 1958 International Geophysical Year. The studies confirmed both the abruptness and timing—about 13,000 years ago—of the Ice Age's ending in Antarctica, but left unexplained the reason for the phenomenon. Additional recent studies support those conclusions: A study of ancient temperatures (*Nature,* 26 February 2009) determined that while warming at the end of the last Ice Age was relatively gradual in Greenland

Figure 43

(north Atlantic), it was "rapid and abrupt" in Antarctica (south Atlantic) about 13,000 years ago. Another study, of ancient sea levels (published in *Science,* 6 February 2009), confirming that Antarctica's ice sheet collapsed abruptly, concluded that due to the topography of the continent and its surrounding sea beds, *the tidal wave was at least three times higher* than hitherto calculated, reaching its maximal impact some 2,000 miles away. A diagram accompanying the article shows the area of maximal tidal impact stretching from the Persian Gulf to the Mediterranean Sea and northeast therefrom—the very Lands of the Bible, all the way to Mount Ararat.

A Deluge date circa 13,000 years ago—at about 10950 B.C.—also dovetails with statements in cuneiform texts that the Deluge occurred in the Age of the Lion; that zodiacal Age indeed began circa 11000 B.C.

Adding 432,000 + 13,000, we can thus confidently say that "Kingship was brought down [to Earth] from the heavens" some 445,000 years ago.

It was then that astronauts from another planet, whom the Sumerians called **Anunnaki**, arrived on Earth. They were the biblical ***Anakim***—the *Nefilim* of Genesis 6.

* * *

The various lists of pre-Diluvial rulers unanimously agree that Eridu was the first city on Earth. The name, ***E.ri.du,*** literally meant 'House in the Faraway Built'; it is a word that has taken root in many languages throughout the ages to denote Earth itself: Earth is called *Erde* in German (from *Erda* in Old High German), *Jordh* in Icelandic, *Jord* in Danish, *Airtha* in Gothic, *Erthe* in Middle English. It was called *Ereds* in Aramaic, *Ertz* in Kurdish—and, to this day, *Eretz* in Hebrew.

It is also important to remember that the various lists of reigns in the initial Cities of the Gods are lists of their successive "chief officers" and not the names of the gods to whom those cities were granted as 'cult centers'. All the lists agree that Alulim and Alalgar were the first rulers in the first city, Eridu; but as is clearly stated in tablet CBS-10673, Eridu was forever granted to **Nudimmud**—an epithet of Ea/Enki that meant 'He Who Fashions Artifacts'; it remained his 'cult center' forever, no matter who was the Chief Administrator ("king") there. (Likewise, Sippar forever remained the 'cult center' of the god Utu, better known by his Akkadian name Shamash; Shuruppak was always linked to the *Sud*—'Medic'—Ninḫarsag; and so on.

Various texts link the establishment of Eridu to the arrival of the Anunnaki on Earth, when 'Kingship' was brought down from heaven.

Just as NASA's first astronauts splashed down in the ocean in their command modules before landing facilities for spacecraft were developed, so did the first group of Anunnaki who came to Earth. They splashed down in the "Lower Sea" (the Persian Gulf), and—dressed in wetsuits, resembling 'Fish-men' (see Fig. 23)—they waded ashore to establish a Home-away-from-home—Eridu—at the edge of the marshlands—a delta formed by the twinlike Tigris and Euphrates Rivers as they flow into the Gulf.

That first group numbered fifty. Their leader, all the texts agree, was **E.A**—'Whose abode is water', the prototype Aquarius. ***'Oannes' had arrived on Earth.***

Several Sumerian texts deal with and describe that First Arrival. One, titled by scholars *The Myth of Enki and the Earth, Enki and the World Order,* or *Enki and the Land's Order,* contains an actual autobiographical account by Ea/Enki. The long text (restored from tablets and fragments scattered between two museums) includes the following first-person statements by him:

> I am the leader of the Anunnaki.
> Engendered by fecund seed,
> the Firstborn son of divine An,
> the 'Big Brother' of all the gods.
>
> When I approached from heaven,
> bountiful rains poured down from the sky.
> When I approached Earth, there was high tide.
> When I approached its green meadows,
> heaps and mounds were piled up at my command.

One of the first tasks was to establish a command post, a headquarters house; it was built at the edge of reed-growing marshes:

> I built my abode in a clean place,
> I called it by a good name,
> its good fortune to portend.
> Its shade stretches over the Snake Marsh,
> its [. . .] has a 'beard' (?) that reaches the [. . .]

Some of the oldest cylinder seals, illustrating Sumer's earliest times, depict reed huts of the kind that the Anunnaki could have erected from the readily available reeds at the edge of the marshes; they all depict inexplicable antenna-like devices protruding from the roofs of those reed huts (Fig. 44).

His outpost needed to be built on an artificial mound, raised higher than the level of the river and marsh waters; Enki assigned the task to one of his lieutenants named Enkimdu:

> After he had cast his eye on that spot,
> Enki raised it above the Euphrates . . .
> Enkimdu, the one of ditch and dyke,
> Enki placed in charge of ditch and dyke.

Enki, the text continues, then gathered his lieutenants at his command post. They included "The weapons-carrying [. . .]," the "Chief-pilots," the "Chief of supplies," the "Lady of grinding and milling," and "the [. . .] who purifies the water." Besides shelter, nourishment had to be found, and the marshes offered an ample fresh supply: "The carp-fish wave their tails among the reeds, the birds chirp to me from their . . .", Enki wrote. Subsequent sections of the text, written in the third person, record Enki's orders to his lieutenants:

> In the marshland
> he marked out a place of carp and fish.
> Enbilulu, Inspector of Canals,
> he placed in charge of the marshlands.
> He marked out a canebreak,

Figure 44

in it he placed [. . .]-reeds and green reeds,
and marked out the cane thicket.

He issued an order to [. . .],
him who set up nets so no fish escape,
whose trap no [. . .] escapes,
whose snare no bird escapes,
[. . .], the son of a god who loves fish,
Enki placed in charge of fish and birds.

The location of those activities is indicated by several references to the two rivers Tigris and Euphrates where they come close to each other, close enough for Enki to make the two meet and cause their "pure waters to eat together."

Several additional sections of the text deal with water-related activities following the arrival. Enki himself is credited with waterworks affecting the two rivers, and other lieutenants are named for water-related tasks: "He filled the Tigris with sparkling water . . . In order to make the Tigris and Euphrates eat together . . . Enki, lord of the deep waters, placed Enbilulu, the Inspector of Canals, in charge thereof." But breaks in the tablets or use of undeciphered terminology leave the nature of some water-related operations uncertain; these include a *sea-water* assignment to a female lieutenant whose epithet-name, **Nin.Sirara** (= 'Lady of Bright Metal'), suggests duties linked to precious metals.

Other unexpected references to metals—*specifically, to gold*—are also made in a section dealing with Enki's post-arrival waterborne inspection of his watery wonderland. He toured the surroundings in a rowboat whose commander held a "rod for [detecting? measuring?] gold" in his hand; his epithet-name, **Nim.gir.sig**, meant "Chief Determinator of Luster." Depictions on early cylinder seals (Fig. 45) show Enki in a reed boat, navigated among the reeds, with a lieutenant-god holding a rodlike device. The boat is equipped at both ends with poles to which are attached circular devices, akin to those placed atop the reed huts.

What do all these tidbits of information mean?

It behooves us to ask at this point a key question regarding the

Figure 45

Anunnaki's coming to Earth: Was it accidental—were they traveling in a spacecraft and, due to a mishap, looked for some solid ground to land on in an emergency, and found the speck of firm ground called 'Earth'? Were they, perhaps, explorers roaming space for pleasure or research, who saw (as Enki described) a watery, verdant place and stopped by to take a look?

In such circumstances, the visit to our planet would have been a one-time event. But the overwhelming ancient evidence indicates that the "visit" lasted an incredibly long time, that it entailed permanent settlements, that the Anunnaki kept coming and going, and that even when a calamity—the Deluge—destroyed all, they stayed on and started all over. This is a pattern of a planned colonization—for a purpose.

Enki and his crew of fifty had come to Earth, I have suggested, for the purpose of obtaining gold.

That purpose emerges, and the tidbits of information begin to make sense, if treated as dots to be connected to what followed. The plan was to extract the gold from the waters of the Persian Gulf. But when this did not work, a change to deep mining had to be undertaken. In that second phase of the Anunnaki's activities on Earth, other gods arrived; leading them was **En.lil** (= 'Lord of the Command') for whom a new city of the gods, ***Nippur,*** was established; its heart was a command and communications hub where orbit-controlling "Tablets of Destinies" spun and hummed in the ***Dur.an.ki,*** the 'Bond Heaven-Earth' Holy of Holies.

While Enlil took charge of the ***E.din*** with its settlements—each with distinct functions—Enki's tasks shifted to a new location called **Ab.zu**; it is a term commonly translated 'The Deep', but which literally read means 'Place of shining metal'.

In *The 12th Planet* I have suggested that the meaning of this combination of two syllables, that in Sumerian could be read in reverse **Zu.ab** without changing its connotation, has been retained in Hebrew as *Za.ab* = '***Gold***'. **Ab.zu/Zu.ab** thus meant the place from whose depths the shining metal—gold—was obtained; the "depth' connotation indicated that the gold was obtained by mining. The Abzu, according to all relevant Sumerian texts, was located in a distant region called ***A.ra.li*** (= 'The place of the shining lodes by the waters') in the "Lower World"; it is a geographic term that applied—in varied texts, including some dealing with the Deluge—to southern Africa. ***Arali,*** I wrote, was in southeastern Africa—a gold-mining region to this day.

The changes that accompanied the second-phase activities of the Anunnaki involved more than a shift from an attempt to easily extract gold from seawater to the need of obtaining it by arduous deep mining. It also involved a change in mission policies, a change of commanders, and the unintended transfer to Earth of personal rivalries and clan clashes from the Anunnaki's home planet, Nibiru, to planet Earth. Varied texts, among them the *Atra-Ḫasis Epic,* detail the tale of the ensuing events; they were, as we shall see, ***the forerunners to the Creation of Man, the explanation for the circumstances of the Deluge, and the key to unlocking the enigmas of the demigods.***

The Arrival, as described in Enki's autobiographical text, was not the Beginning of the chain of momentous events; for the Real Beginning we have to start with the tale of Creation itself, just as the people of Mesopotamia had done each New Year time. We have to read, re-read, and understand the Mesopotamian *Enuma elish* and the biblical tale of Genesis. The precise information they provide not only explains many of the phenomena in our Solar System and beyond—it sheds light on the Origins of Life, on Who we are, and How we came to be here, on Planet Earth.

ABODE OF GOLD AND FLOWING WATERS

The Abzu, a Sumerian text stated, was in the ***Ut.tu,*** "in the west"—west of Sumer (as southeast Africa is), reachable by seafaring boats that traversed "the distant sea." Its ***Arali,*** its mining zone, was reported in one text to have been "120 ***beru*** of waters away from the quay on the Euphrates"—a distance of traveling 120 "double hours"—ten days of traveling by sea. The mined ores were transported to the *Edin* by cargo ships called ***Magur Urnu Abzu*** (= 'ship for ore of Lower World'), for smelting, refining, and formation into portable ingots called **Zag** (= 'Purified Precious').

A Sumerian hymn in praise of the Abzu described the place where Enki established his new headquarters as one with water rapids or great waterfalls:

> To thee, Abzu, pure land
> where great waters rapidly flow,
> To the 'Abode of Flowing Waters'
> the lord [Enki] betook himself.
>
> The 'Abode of Flowing Waters'
> Enki among the pure waters established.
> In the midst of the Abzu
> a great sanctuary he established.

Sumerian-Akkadian syllabaries stated that "Abzu = *Nikbu*"—a deep and tunneled mine. The initial pictogram for ***Abzu*** (from which its cuneiform sign evolved) represented a mining shaft—variants of which stood for gold and other mined minerals, including precious stones:

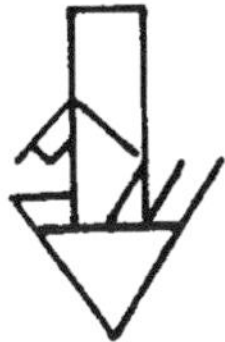  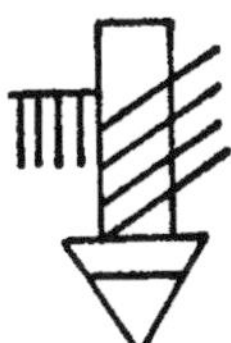   

Once the obtainment of gold from the Abzu was in full swing, Enki's autobiographical text extolled the region as ***Meluhha***, "Black land of large trees . . . whose laden boats transport gold and silver." Later Assyrian inscriptions identified Meluhha as *Kush*, "land of the dark-skinned people" (= Ethiopia/Nubia). The syllabic components of the Sumerian term convey the meaning 'Fish-filled Waters'; the term may thus explain cylinder seal depictions of fish-filled waters flowing from Enki, who is flanked by workers holding typical gold ingots:

VI
A Planet Called 'Nibiru'

The notion of space travel is no longer relegated to science-fiction alone. Serious scientists do not rule out that one day, some day, we Earthlings might send astronauts not just to our celestial satellite the Moon, but also to another planet farther out. Some savants even dare acknowledge that life, even like ours, might exist 'somewhere' in the vast universe with its countless galaxies and constellations and billions of stars ('suns') orbited by satellites called 'planets'. But such sentient beings, even if clever enough to have their own space program—so the argument goes—could never visit us (or we them) because the nearest possible place in the heavens where such life could exist is "light-years away"—a Light-Year being the ungraspable distance that light travels in one year.

But what if such a compatible planet were to exist much closer—say, in our own solar system? What if travel between it and Earth needs only so many 'normal' years, not Light-Years?

That is not a theoretical question, because that is precisely what we are told by the ancient texts—if only we stop treating them as myth and fantasy and consider them to be factual recollections and records of actual events. It was by doing that, that the trailblazing *The 12th Planet* book became possible.

Logically, for Eridu in Mesopotamia to be 'Home away from home', there had to be a Home from which Enki had come. For his crew of fifty to be called "Those who from Heaven to Earth came" (= **Anunnaki**), they had to come from a place, an actual place, in the heavens. Thus, there had to be a place, somewhere in the heavens, where the journey to Earth began—a place where intelligent beings, capable of space travel some 450,000 years ago, could live. We can call it 'Planet X' or 'Planet of the Anunnaki'; in ancient Mesopotamia it was called ***Nibiru;*** its ubiquitous symbol throughout the ancient world was the Winged Disc (see Fig. 10); its orbit was traced and observed most reverently; ***and it is beyond dispute that countless texts, starting with the Epic of Creation, refer to it by name repeatedly.***

When, at the end of the 19th century, astronomical tablets from Mesopotamia were found and deciphered, the savants at the time—Franz Kugler and Ernst Weidner stand out to this day—debated whether ***Nibiru*** was just another name for Mars or for Jupiter; it was an accepted axiom that the ancients could not be aware of any planet beyond Saturn. *It was a major breakthrough moment when it dawned on me, in the middle of one night, that Nibiru is neither Mars nor Jupiter—* ***that it is the name of one more planet in our own Solar System.***

One can start the chain of evidence where the Hebrew Bible has it, in verse 1 of chapter 1 of Genesis: "*In the beginning, God created* the Heaven and the Earth." So begin virtually all the translations of the Hebrew Bible's first three words, *Bereshit bara Elohim* (for the moment, we shall treat such a translation as valid). Continuing with just 31 verses, the Hebrew Bible then encapsulates Creation, from how the Sky above with the "Hammered Bracelet" and the Earth below were formed, to how life on Earth came to be—from grasses to marine to vertebrate and mammalian, and finally Man. The biblical sequence (including a dinosaur phase, in verse 21) matches modern scientific findings about Evolution, so that the notion that Bible and Science are in conflict is baseless.

The discoveries of the inscribed tablets of the Mesopotamian 'Epic of Creation' (as described in a previous chapter) leave no doubt that whoever had written the biblical rendering was well aware of the tale in *Enuma elish,* condensing its six tablets plus a laudatory seventh to six phases ("days") of Creation plus a sanctified seventh "day" of divine gratification.

Such awareness of the sequence rendered in *Enuma elish* was not only possible due to the proliferation and durability of the tablets containing the text; it was probably unavoidable, because the Epic of Creation was read in public as part of the annual New Year festival, first in Sumer and then in Babylon, Assyria, and beyond—throughout the ancient Near East. The reading started at evetime on the festival's fourth day, and lasted through the night, for *Enuma elish* (as the most complete Babylonian version of the epic is titled) is long and detailed. Its central religious-scientific aspect was a battle between a celestial goddess called "Tiamat" and a heavenly Avenger-cum-Savior god—the main reason why the text has been treated by modern scholars both as a myth and as an allegorical tale of Good vs. Evil, a kind of ancient 'St. George and the Dragon' tale.

In *The 12th Planet* I audaciously suggested instead that the Epic of Creation is at its core a great scientific text that starts with a cosmogony that embraces the whole Solar System, explains the origins of Earth, Moon, and the Asteroid Belt, reveals the existence of planet Nibiru, proceeds through the arrival of the Anunnaki gods on Earth, and describes the creation of Man and the rise of civilization; adapted to promote religious-political purposes, an appended ending hails the victorious assumption of supremacy by the relevant national god (**Enlil** in Sumer, **Marduk** in Babylon, **Ashur** in Assyria).

Irrespective of version, when the primeval events began, "Heaven above," and "firm Earth below" had yet to come into existence:

> *Enuma elish la nabu shamamu*
> When in the above Heaven had not been named
> *Shaplitu ammatum shuma la zakrat*
> [And] below Firm Earth had not been called—

—at that primeval time, the ancient text states, the Solar System began to take shape with just three celestial actors: A primordial **Apsu**, its companion **Mummu**, and a female celestial entity called **Ti.amat.** (The three names in the Babylonian text have been retained unchanged from the undiscovered Sumerian original, and mean, respectively, 'One who exists from the beginning', 'One who was born', and 'Maiden who gives life'.)

Celestial gods—planets—then begin to be engendered as Tiamat, a watery planet, starts to "mingle the waters" with the male Apsu (the Sun). First, the pair **Laẖamu** and **Laẖmu** are formed in the space between them; then—"surpassing them in stature"—the larger pair **Kishar** and **Anshar** appear; and finally, the pair **Anu** and **Nudimmud** are formed farther out. These are Sumerian names (attesting to the Sumerian origin of the epic), except that *Anu* is Babylonian for the Sumerian **An** (= 'The Heavenly One').

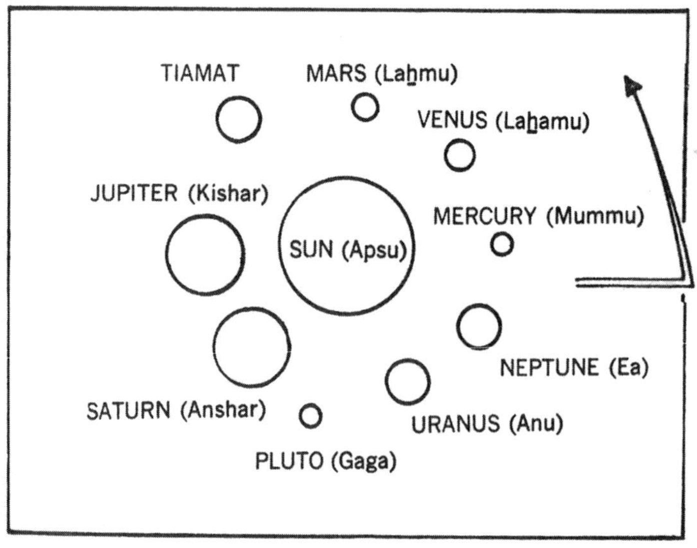

Figure 46

The resulting Solar System (Fig. 46) accurately conforms to *our solar system* and its planetary layout as we know them (except for Tiamat, of which much more soon):

> SUN—Apsu, "One who existed from the beginning."
> MERCURY—Mummu, "The one who was born," the Sun's companion.
> VENUS—Laẖamu, "Lady of battles."

MARS—La<u>h</u>mu, "Deity of war."
—??—**Tiamat**, "Maiden who gives life."
JUPITER—Kishar, "Foremost of firm lands."
SATURN—Anshar, "Foremost of the heavens."
Gaga—Anshar's messenger, the future PLUTO
URANUS—Anu, "He of the heavens."
NEPTUNE—Ea/Nudimmud, "Artful creator."

Modern science holds that our Solar System was formed about 4.5 billion years ago when a whirling cloud of cosmic dust ringing the Sun began to coalesce, forming planets orbiting it—planets spaced out in the same orbital plane (called Ecliptic) and circling in the same direction (counterclockwise). The description in the ancient Mesopotamian epic is in accord with these modern findings but offers a different (and probably more accurate!) *sequence* of planet formation. The Sumerian names of the planets are meaningful and accurate descriptions of these celestial bodies—facts that modern astronomy keeps discovering, as for example a 2009 discovery that it is indeed Saturn ('Anshar') and not the more massive Jupiter ('Kishar') that is "Foremost of the heavens" due to its system of rings that tremendously extend its reach.

The resulting Solar System, the epic relates, was unstable and chaotic at its start. The planetary orbits were not yet firmly set: "The divine brothers banded together"—getting in each other's way. "They disturbed Tiamat as they surged back and forth"—moving in unstable orbits, crowding toward Tiamat. Even the gravitational and magnetic forces of the Sun were ineffective—"Apsu could not lessen their clamor." Again, modern science too, discarding a longheld notion that once the Solar System was formed it was done, now finds that it was unstable for a long time after its formation, and that shiftings and collisions were taking place.

The unstable celestial gods, "by their antics in heaven," were now "troubling the belly of Tiamat," *Enuma elish* relates. They were causing her to sprout her own fearsome "assembly"—a group of her own satellite moons. This, in turn, brings more turmoil that endangers the other celestial gods. At this dangerous phase, the outermost celestial god

Nudimmud (our Neptune) takes matters into his hands: "Surpassing in wisdom, accomplished, resourceful," this celestial deity balances the wobbly Solar System by inviting in an outsider—one more large celestial god.

The newcomer was not formed with the others; it is a stranger coming from afar. It originated far out "in the heart of the Deep," and it is "filled with awesomeness"—

> Alluring was his figure,
> Flashing [was] the gaze of his eyes.
> Lordly was his gait,
> Commanding from the beginning.
> Artfully arranged, beyond comprehension,
> were his members—
> beyond understanding, difficult to look upon . . .

Subjected to the gravitational pull of 'Nudimmud' and coming under the influence of the other planets, the stranger from outer space curves its course toward the Solar System's center (Fig. 47). When it passes too

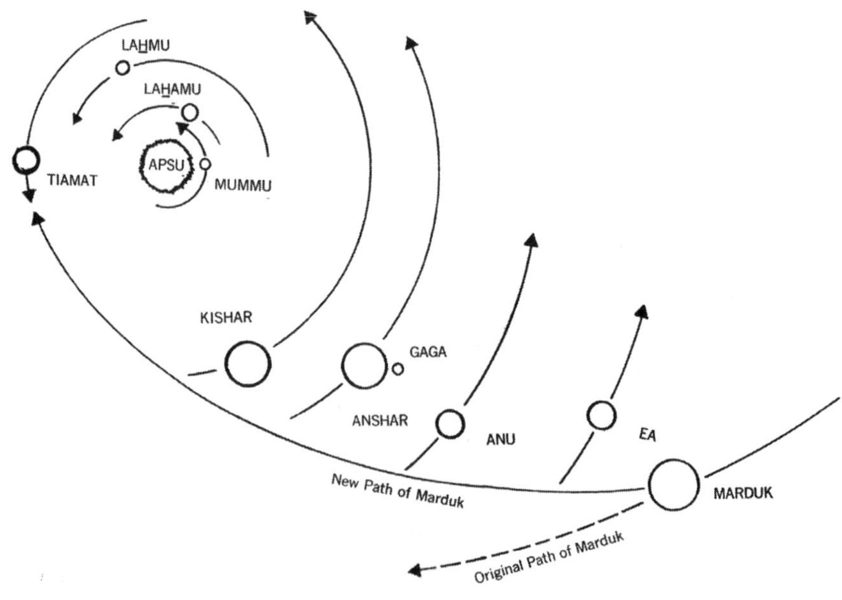

Figure 47

closely near Anu (our Uranus), the cumulative gravitational forces tear off it chunks of matter and the invader sprouts four "winds"—satellites, moons—that whirl around it.

One cannot be certain whether the original Sumerian text had already named this stranger from outer space ***'Nibiru'*** at this point; but it is certain that the Babylonian version changed it here to **Marduk**—the name of Babylon's national god. This transformation of Marduk from a god on Earth to a celestial deity by renaming Nibiru 'Marduk', was accompanied in the Babylonian text by the revelation that 'Nudimmud'—who "engendered" the newcomer by inviting him in—is none other than Ea/Enki, the real father of the Babylonians' god Marduk, and that Anu is Ea/Enki's father (as proclaimed, in fact, by Enki in his earlier-quoted autobiography). Thus, with a sleight of hand, the celestial tale became a religious-political legitimization of a dynasty: Anu > Ea/Enki > Marduk . . .

As the ancient text describes the progression of the invading planet, it becomes clear that it is ***moving in a clockwise direction***—the opposite or 'retrograde' of the other planets' counterclockwise orbital direction. *It is a finding that offers the only explanation to varied otherwise inexplicable phenomena in our Solar System.*

This 'retrograde' direction of Nibiru's path made an eventual collision with Tiamat inevitable; and the ensuing "Celestial Battle," as the ancient text calls the collision, was a basic tenet of ancient knowledge, reflected in countless references in the Bible's books of Psalms, Job, and the Prophets.

Disturbed by the new gravitational forces, "pacing about distraught," Tiamat gives rise to her own defensive host of eleven satellite-moons; the Babylonian text describes them as "roaring dragons, clothed with terror." The largest of them, **Kingu**, is Chief of her host: "Kingu she exalted, in their midst she made him great"; his task is to prepare for battle with the oncoming Marduk. As Kingu's reward, Tiamat readies him to join the "Assembly of the gods"—*to become a planet in his own right*—by granting him a Celestial "Destiny" (an orbital path). That alone was reason enough for the Sumerians (and their successors) to count this particular Moon as a member in its own right of our Solar System.

As the stage is set for the Celestial Battle, Tablet I of *Enuma elish* comes to an end; and the scribe of the best preserved version, one Nabu-mushetiq-umi, inscribes at its end the customary colophon: "First tablet of Enuma elish, like the original tablet [. . .], a copy from Babylon." He also identifies the scribe whose tablet he copied—a tablet "Written and collated by Nabu-balatsu-iqbi, the son of Na'id-Marduk." The copying scribe then dated his work: "The month of Iyyar, the ninth day, the twenty seventh year of Darius."

Discovered at Kish, this first tablet of *Enuma elish* is thus identified by its scribe as a copy made at the start of the 5th century B.C. during the reign of Darius I. By a twist of fate, it was the same Darius whose rock inscription in Behistun (see Fig. 17) enabled Rawlinson to crack open the mystery of cuneiform writing.

* * *

Tablet II of *Enuma elish* tracks the emergence of two opposing planetary camps headed for the inevitable collision.

Treating the celestial gods as living entities, the text tells that while Tiamat was forming her ferociously whirling satellites, in the Solar System's outer reaches Ea/Enki appealed to his 'grandfather' Anshar to organize the varied planets and have them anoint 'Marduk' as their leader in battling Tiamat and her host: "Let him who is potent be our Avenger, let Marduk, keen in battle, be the hero!"

A crucial stage is reached when 'Marduk' nears the colossal Anshar, for Anshar (Saturn) has "lips"—majestic rings—that extend out off the face of Anshar. Encountering them, the approaching Marduk "kisses the *lips* of Anshar" (the *rings* of Saturn). The passage by and 'acceptance' by the dynastic great-grandfather encourages Marduk to voice his wishes: "If I, indeed, as your Avenger, am to vanquish Tiamat . . . convene an Assembly to proclaim my destiny supreme!" A celestial "destiny"—an orbit—greater than that of all the other planets is Marduk's demand.

It is here (by now Tablet III) that, according to Sumerian cosmogony, *the future Pluto obtains its planetary status and unique orbit.* A moon of Anshar/Saturn called ***Gaga,*** it is detached by the force of the oncoming Marduk and is thrust out as an emissary to Laẖmu and Laẖamu, purportedly to canvass their vote for the elevation of Marduk to leadership. When Gaga returns, it circles back all the way to the outermost Ea/Neptune; there it becomes the planet we call Pluto with its oddly inclined orbit that takes it at times beyond and sometimes inside the orbit of Neptune. (Aware of that unusual orbit, the Sumerians depicted the planet as a two-faced deity, seeing its master Ea/Enki/Neptune once this way and once the other way, Fig. 48.)

With all the planets opposing Tiamat agreeing to Marduk's demands for supremacy (Tablet IV), the giant Kishar/Jupiter adds more weapons to Marduk's arsenal: In addition to the four satellites (named "South Wind, North Wind, East Wind, West Wind") that he had obtained from Anu/Uranus, three new awesome satellites ("Evil Wind, Whirlwind, Matchless Wind") are added, creating an awesome whirling battle entourage of "seven in all."

Thus reinforced, Marduk—"filled with a blazing flame," able to shoot lightnings as arrows, possessing a magnetic field to "ensnare Tiamat as in a net"—"Toward the raging Tiamat set his face." Tiamat,

Figure 48

meanwhile, is orbiting in a direction toward the oncoming Nibiru/ Marduk; the Celestial Battle, the collision, was about to occur:

> Tiamat and Marduk, the wisest of gods,
> advanced against each other.
> They pressed on in single combat,
> they approached for battle.
>
> The four winds he stationed
> that nothing of her could escape:
> The South Wind, the North Wind,
> the East Wind, the West Wind.
>
> Close to his side he held the net,
> gift of his grandfather Anu.
> He brought forth the Evil Wind, the Whirlwind
> and the Hurricane to trouble Tiamat's insides.
> All seven of them rose up behind him.
>
> In front of him he set the lightning,
> with a blazing flame he filled his body,
> With a fearsome halo his head was turbaned,
> he was wrapped with awesome terror as with a cloak.

As the two hurtling planets neared each other, Marduk went on the attack:

> The Lord spread out his net to ensnare her;
> The Evil Wind, the rearmost, he let loose in her face.
> When Tiamat opened her mouth to devour it,
> he drove in the Evil Wind, that she close not her lips.

Tiamat, according to this step-by-step account of the battle, was first struck with one of Marduk's seven satellites where her 'mouth' was. Then Marduk's other moons served as weapons:

> The raging Winds then charged her belly;
> Her belly was distended, her mouth was opened wide.

> He shot through it an arrow, it tore her belly.
> It cut through her insides, fracturing her midst.
> Having thus subdued her, he extinguished her life.

So, according to Sumerian cosmogony as retained in *Enuma elish,* in this first encounter between Marduk and Tiamat, the two planets did not collide: It was the "winds"—satellites—of Marduk that struck Tiamat, "fracturing her midst" and "extinguishing her life." We illustrate that first encounter in Fig. 49.

While the final blow to the gashed Tiamat is yet to be delivered in a subsequent encounter, in this first round Marduk and his Winds deal with Tiamat's "host" of orbiting satellites. The smaller ones, "shattered, trembling with fear, turned their backs about to save their lives . . . tightly encircled, they could not escape." The phrase *"turned their*

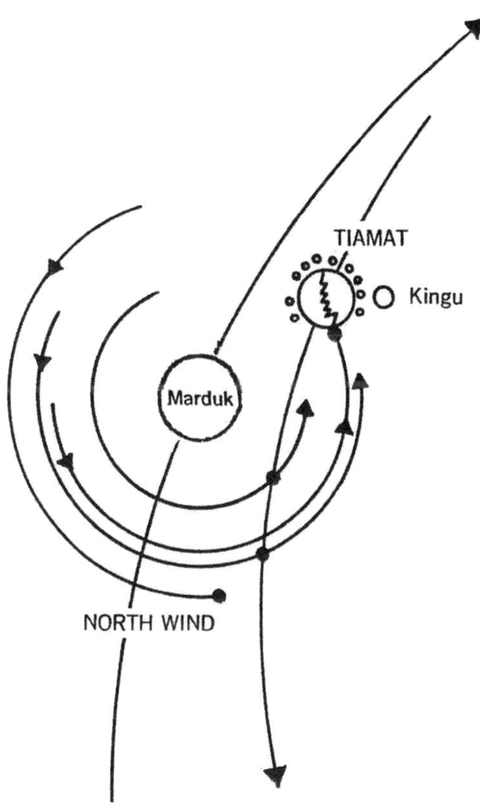

Figure 49

backs about"—thrust in the direction of the advancing Marduk—they become the otherwise inexplicable *retrograde orbiting* comets.

Kingu, their leader, "rendered lifeless," is bound and held captive; he is deprived of the "Tablet of Destinies" that was about to make it a planet in its own right. Snatching it, Marduk "took away from him the Tablet of Destinies, not rightfully his," and transferred the orbital capability to himself. Devoid of an atmosphere, Kingu is turned into a ***Dug.ga.e,*** a Sumerian term that can best be translated as "Lifeless Circler"—doomed forever to keep circling Earth.

Now enabled to go into orbit, Marduk circles back to revisit Anshar and Ea/Nudimmud and reports his victory to them. As he completes his first solar orbit, he is coming back to the site of the Celestial Battle: Marduk "turned back to Tiamat, whom he had subdued." ***This time, Marduk himself collides with the wounded Tiamat, cleaving her apart:***

> The Lord paused to view her lifeless body.
> To cleave the monster he then artfully planned.
> Then, as a mussel, he split her into two parts.

The fate of the two parts is of crucial importance; every word in the ancient text is significant, for it is here that we are witnessing the Anunnaki's sophisticated understanding of how Earth, the Moon, and the Asteroid Belt came to be:

> The Lord trod on Tiamat's hinder part.
> With his weapon her skull he cut loose;
> The arteries of her blood he severed,
> and caused the North Wind to bear it
> to places that have been unknown.
>
> The [other] half of her
> as a screen for the skies he set up.
> He bent Tiamat's tail,
> as a bracelet the Great Band to form;
> Locking the pieces together,
> as watchmen he stationed them.

In *The 12th Planet* I have suggested that ***the severed upper half ("skull") of Tiamat, thrust off to another place in the Solar System, became the planet Earth*** in a new orbital path; that Kingu, doomed to become a "Lifeless Circler," was carried with it to become Earth's Moon; and that ***the hinder part of Tiamat, smashed to bits and pieces, became the Asteroid Belt (the "Great Band" or "Hammered Bracelet")***—Fig. 50. That the shattered smaller moons of Tiamat became the puzzling retrograde comets that "turned back" and assumed Marduk's retrograde orbit is reinforced by the statement that 'Marduk' "tied them to his tail"—pulling them in his own retrograde orbital direction.

This understanding of the Creation tale, reaffirmed repeatedly in various Sumerian texts, also offers the only plausible explanation for the biblical verses in Genesis dealing with the event—and the origin of life on Earth:

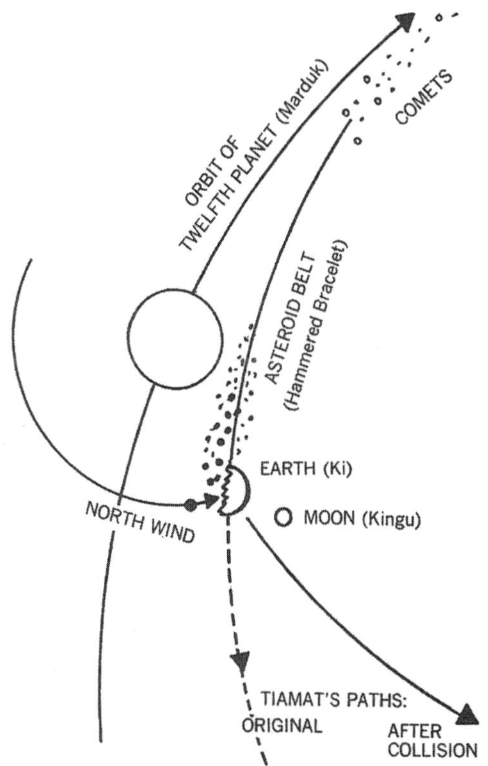

Figure 50

- In the first encounter, satellite/moons of 'Marduk' strike and disable Tiamat
- In the second decisive encounter, 'Marduk' itself "treads upon"—strikes and *comes in contact* with—Tiamat, splitting her in two; it is thus that *the "seed of life" present on Marduk is transferred to and shared with the future Earth.* Keeping Tiamat's waters, it is a future watery planet
- The upper half ("skull") of Tiamat is thrust off to a new orbital location to become the Earth, now seeded with DNA from Marduk
- The thrust half (the future Earth) carries with it the lifeless Kingu to become its Moon
- The bottom part is smashed to bits and pieces; tied together as a bracelet, it becomes the Asteroid Belt
- Where the Celestial Battle had taken place, where Tiamat had once orbited, is termed *Shamamu* in Akkadian, and *Shamay'im* in Hebrew—terms that are translated as '***Heaven***' but which stem from *Ma'yim,* "waters"—the place where the watery Tiamat used to be.

In the Mesopotamian texts, the affirmation of this sequence was repeatedly expressed by the following statement:

> After Heaven had been ***separated*** from Earth,
> After Earth had been ***moved away*** from Heaven

* * *

Having reshaped the heavens, created Earth, and fashioned the Hammered Bracelet, Marduk "crossed the heavens and surveyed the regions . . . his Great Abode he measured." Liking what he saw, the Mesopotamian text states, "He (Marduk) founded the station of Nibiru."

Celestially, by making our Solar System his abode, ***'Marduk' has become Planet Nibiru.*** A tenth planet, a twelfth member of the Solar System (Sun, Moon, and ten planets) has been added—exactly as is depicted on a cylinder seal from 2500 B.C. (cataloged VA-243 in the

Vorderasiatisches Museum in Berlin, Fig. 51, with enlarged sketch added). The similarity to the order of planetary formation per *Enuma elish* (depicted in Fig. 46), speaks for itself.

The new planet's orbit stretched from "the Apsu's region to the abode of Ea"—from an "abode" (Perigee) near the Sun to an "abode" (Apogee) well beyond Neptune (Fig. 52). With this great elliptical orbit,

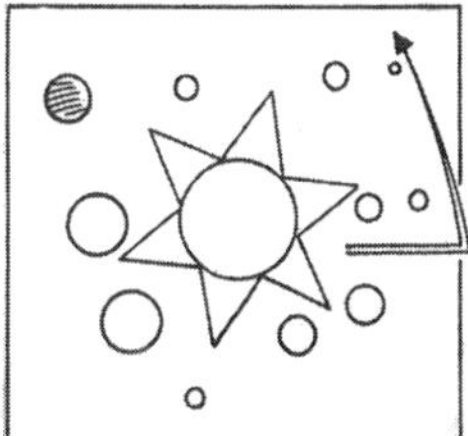

Figure 51

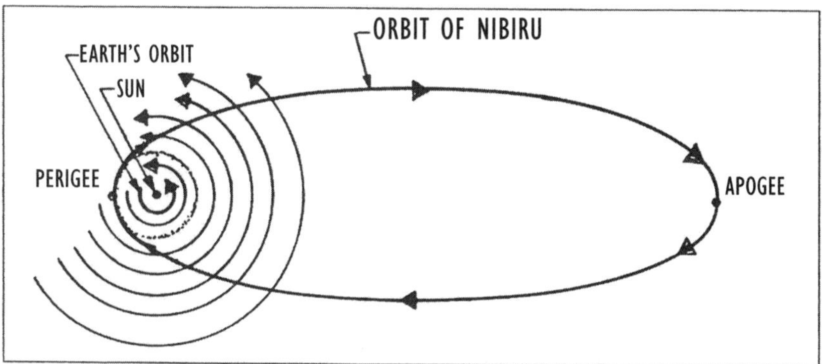

Figure 52

Marduk's celestial "destiny" became supreme—just as he had been promised.

This orbit, the epic states, is what gave the new member of our Solar System its name, for ***Nibiru*** means 'Crossing':

> Planet Nibiru:
> The crossroads of Heaven and Earth he shall occupy.
> Above and below [the gods] shall not go across;
> They must await him.
>
> Planet Nibiru:
> Planet which is brilliant in the heavens.
> He holds the central position;
> To him the gods shall pay homage.
>
> Planet Nibiru:
> It is he who without tiring
> the midst of Tiamat keeps crossing.
> *Let 'Crossing' be his name!*

Called *Shar* (= 'The King's'), this orbit equated mathematically 3,600, suggesting that this was the orbital period of Marduk/Nibiru—3,600 *Earth*-years. As it returns annually (one orbit being one year for Nibiru!) to its perigee, where Tiamat had been, Nibiru *intersects the ecliptic;* it is its *Crossing Point;* and whenever Mankind had witnessed that occurrence, Nibiru was depicted as a radiating planet symbolized by the sign of the Cross (Fig. 53).

Geological, geophysical, and biological evidence gathered on Earth, on the Moon, and from asteroids and meteorites, has convinced modern scientists that a cataclysm, a "catastrophic collision event" affecting our part of the Solar System, had occurred circa 3.9 billion years ago—about 600,000 years after the formation of the Solar System. The "event," I have suggested, was the Celestial Battle between 'Marduk' and Tiamat.

* * *

Figure 53

Enuma elish filled up four tablets with the Tale of Creation thus far; the Hebrew Bible did it in eight verses and two Divine Days.

In the familiar King James translation, we learn (verses 1–5) that when the creation of Heaven and Earth began, the Earth "was without form and void" and "the Deep" was in darkness. Then "the Spirit of God moved upon the waters"; and God commanded "Let there be light, and there was light." And having "divided the light from the darkness," God "called the light Day and the darkness Night"; and "it was evening and it was morning, Day One."

One would be less hard put to discern in those words their Mesopotamian origin if the *actual Hebrew text* is followed. There, the darkness was not "upon the face of the Deep" but upon *Tehom* (Hebrew for ***Tiamat***); it is *Ru'ah̲* (***wind,*** not "spirit"), Marduk's satellite—that moved against *Tehom*/Tiamat, as his ***lightning,*** not mere "light," struck her.

Verses 6–8—the events of Day Two—translations use the term "Firmament" (to describe the Asteroid Belt) where the Hebrew says

Raki'a (*Rakish* in the Babylonian text), which literally means 'Hammered Bracelet'. Located "in the midst of the waters" ***to separate*** the "waters above" from the "waters below," it is the ***Sham-Mayim*** (= 'Place of the waters') that is translated '***Heaven***'.

Choosing to skip the polytheistic sections about the multiple gods' genealogy, rivalries, and discussions, the editor-author of Genesis just restated *the scientific fact* of an Earth cleaved off Tiamat as a result of a celestial collision. The ancient view was that the Hammered Bracelet/ Asteroid Belt served as a "Firmament" or a "Heaven" separating celestial regions; the Hebrew term for that region, the *Shama'yim,* and its meaning, "Heaven," were obviously *borrowed directly* from the opening verse of *Enuma elish: "**elish,** la nabu **shamamu**"*—"in the **Above**, ***Heaven*** had not been named." Indeed, the whole biblical notion of a celestial 'Above' and a celestial 'Below' stemmed from the two opening verses of *Enuma elish:* The 'Above' from the first verse just quoted, and the 'Below' from the second verse: ***Shaplitu,*** *ammatum shuma la zakrat*— "***Below*** firm *Earth* had not been called."

Such a celestial division to an "Above" the Firmament/Heaven and to a "Below" it seem baffling at first glance; but they become pertinent and clear when we illustrate the statement about Nibiru's attaining the Crossing "in the midst" of where Tiamat had been:

Nibiru

Mercury Venus Earth Moon Mars >   < Jupiter Saturn Uranus Neptune Pluto

Asteroid Belt

Passing at its perigee between Mars and Jupiter, Nibiru indeed makes the Crossing in the midpoint between all the other planets of the Solar System (Moon included). As the Bible's terminology explains, the *Shama'yim* (literally, 'Place of the Waters' but translated "Heaven")—the place of the "Firmament" (*Raki'a, Rakish*). The place where Nibiru "crosses" indeed *divides the planetary system* into an "Above" and a "Below"—into the Solar System's Outer Planets in the "Above" and the Inner Planets in the "Below" nearer the Sun.

What *Enuma elish* and the Bible say is confirmed by modern astronomy that refers to the "below" group as the 'Terrestrial Planets' and the "above" group as the "Outer Planets"—separated by the Asteroid Belt.

That basic tenet of ancient cosmology and astronomy is even confirmed by a depiction on a Sumerian cylinder seal, now on view in the Bible Lands Museum in Jerusalem, Israel, that graphically expresses this celestial division (Fig. 54). It uses the straw used in beer drinking as the dividing Asteroid Belt; to its left side the "Below" planets (starting with Venus as the eighth planet, then Earth and its crescent Moon, and Mars nearest the Belt); and on the other side, it shows the "Above" Jupiter and Saturn *with its rings*.

* * *

As Tablet V begins, the continuing *Enuma elish* text then ascribes to Marduk the establishment of "the precincts of night and day" by assigning Night to the Moon, Daytime to the Sun, and credits him with all the Sumerian astronomical achievements: It was he who instituted a luni-solar calendar, fixed the Zenith, divided the heavens into three zones, and grouped the stars in twelve zodiacal constellation, giving them their "images."

We find this segment repeated, almost verbatim, in Genesis 1:14–19, where God is credited with "dividing the day from the night," making the Sun and the Moon responsible "for seasons, days, and years," and "forming the constellations and also their signs."

With all celestial matters taken care of, divine attention shifted

Figure 54

to Earth itself, to making it habitable. In the Mesopotamian text, we reached Tablet V, a complete and almost intact tablet (some 22 lines are still missing) was found only in the late 1950s at an unlikely Turkish site called Sultantepe. From it one learns that after Marduk had given the Sun and the Moon their appointed tasks etc., he turned his attention and creative energy to making Earth—the former upper part of Tiamat—a viable place:

> Taking the spittle of Tiamat
> Marduk created [. . .];
> He formed the clouds, filled them with [water],
> raising the winds for bringing rain and cold.
>
> Putting Tiamat's head into position,
> he formed thereon the mountains.
> He caused the Euphrates and Tigris
> to flow from her eyes.
>
> Stopping her nostrils, he [. . .].
> In her udder he formed the lofty mountains,
> [Therein] he drilled the springs,
> for wells to carry away the [waters].

Clearly, having just been cleaved off from Tiamat, Earth is in need of reworking and reshaping by its creator to become a habitable planet with mountains, rivers, flowing waters, etc. (the "spittle," I suggest, refers to volcanically ejected lava).

Returning to the Bible, we find that Genesis too reports that having completed the celestial arrangements, divine attention turned to Earth. Verses 9–10 describe the steps taken to make it habitable:

> And God said:
> Let the waters under the heaven
> be gathered together unto one place,
> and let the dry land appear;
> and it was so.

> And God called the dry land 'Earth',
> and called the gathered together waters 'Seas'.

This biblical account is in full accord with modern findings that all of Earth's dry land began as one super-continent ('*Pangaea*') that emerged when all the Earth's waters were gathered into one vast '*Pan-ocean*'. Pangaea in time broke up and its parts drifted off away from each other, becoming several continents (Fig. 55). This modern 'Continental Drift' theory is fundamental to all Earth sciences, and to find it clearly stated in the Bible (and probably in the missing lines of Tablet V) is quite remarkable.

The Hebrew and the Babylonian texts provide here a logical and

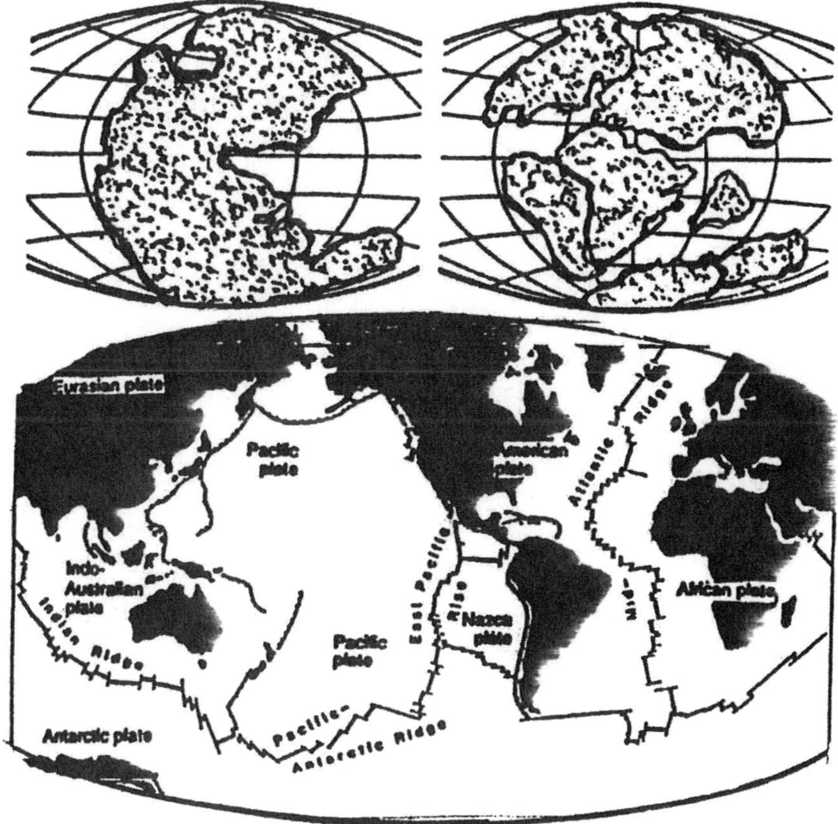

Figure 55

scientifically accurate process: The wounded segment of the watery Tiamat begins to assume a planetary shape; the waters collect in the cavitous part (of which the Pacific Ocean is the largest and deepest), revealing dry land; the continents appear, mountains are pushed up; volcanoes spout lava and gases, giving rise to an atmosphere; clouds and rains come; rivers begin to flow. *Earth is ready for Life.*

"Thus," states *Enuma elish* in Tablet V, line 65, "he (Marduk) created Heaven and Earth."

"Thus," states the Bible in Genesis 2, verse 1, "were completed the Heavens and the Earth and all of their host."

By treating *Enuma elish* as a sophisticated cosmogony and not as an allegorical tale of a struggle between good (the Lord/Marduk) and evil (the Monster/Tiamat), we have obtained a coherent explanation for many puzzles in our Solar System and explain the incredibly fast appearance of life on Earth—and the compatability between the Anunnaki and the Daughters of Man. ***The Bible, I suggest, has done the same.***

THE BEROSSUS VERSION

One must assume that among the crucial texts that were copied and recopied, a version of the tale of Marduk, Tiamat, and the Celestial battle had to be in the hands of Berossus when he compiled his three volume *Babyloniaca.*

This, apparently, he did. According to the historian Alexander Polyhistor—one of the sources for the Berossus Fragments—in Book I, Berossos wrote (among other things):

> There was a time in which there was nothing but darkness and an Abyss of waters, wherein resided most hideous creatures . . .
>
> The one who presided over them was a female named Thallath, which according to Chaldean means "the Sea" . . .
>
> Belus (= 'The Lord') came, and cut the female asunder;
> and out of one half of her he formed the Earth,
> and of the other half the Heavens;
> at the same time he destroyed the creatures of the Abyss . . .
>
> This Belus, whom men call Deus, divided the darkness,
> and separated the Heavens from the Earth,
> and put order in the universe . . .
> He also formed the stars, and the Sun, and the Moon,
> together with the five planets.

Did Berossus have access to a complete and undamaged copy of Tablet V of *Enuma elish?* This interesting question leads to a more general one: Where, in which library, among which collection of tablets, did Berossus sit, copy from the tablets, and write his three volumes?

The answer might lie in the discovery in the 1950s that a mound called Sultantepe, a few miles norh of Harran (now in Turkey), was actually the site of *a major scribal school and library*—where many tablets until then missing were found.

VII
Of Anunnaki and Igigi

It was probably near midnight when the public reading of *Enuma elish* (in Babylon accompanied by reenactments, a kind of Passion Play) had reached the statement that the creation of the heavens and the Earth—by Marduk—has been accomplished. Now it was time to translate his celestial supremacy to supremacy among the Anunnaki—the heavenly gods who came to Earth.

With admirable subtlety, the name **Enlil**—the deity who probably was the hero of the creation tale in its Sumerian original—is mentioned (for the first time) alongside those of Anu and Ea/Enki: It is slipped in in the very last line of Tablet IV. Then, as the tale continues on Tablet V, other deities—including Marduk's real mother, **Damkina** (renamed **Ninki** after Ea was titled 'Enki' = 'Lord of Earth')—take the stage; and the listener (or reader) finds himself witnessing Marduk's coronation as 'king' not only by the Anunnaki gods, but also by another group of deities called **Igi.gi** (= 'Those Who Observe and See').

It is a grand assembly of all the leading gods. Marduk is seated on a throne, and his proud parents, Ea/Enki and Damkina, "opened their mouths to address the great gods," saying thus: "Formerly, Marduk was [merely] our beloved son; now he is your king; proclaim his title "King of the gods of Heaven and Earth!" Compliance to that request/demand followed:

> Being assembled, all the Igigi bowed down;
> Everyone of the Anunnaki kissed his (Marduk's) feet.

They were assembled to do obediance;
They stood before him, bowed, and said:
"He is the king!"

They gave sovereignty to Marduk;
They declared for him a formula of
good fortune and success, [saying]:
"Whatever you command we will do!"

The text does not state where the Assembly is gathered. The narrative suggests that the coronation of Marduk is taking place on Nibiru, and it is followed by an assembly of the gods assigned to Earth. Reminding the gathered gods of his royal lineage (some ancestors who preceded Ea and Anu are invoked), Marduk, as the newly elected Chief, loses no time in outlining his divine program: Hitherto, he tells the gathered gods, you have resided in ***E.sharra,*** "the Great Abode" of Anu on Nibiru; now you will reside in "a counterpart abode thereof that I will build in the Below." "In the Below"—on Earth—Marduk says, he has created Firm Ground suitable for a New Home:

I have hardened the ground for a building site,
to build a home, my Luxurious Abode.
I will establish therein my temple,
its shrines will affirm my sovereignty . . .
I will call its name *Bab-ili* ['Gateway of the gods'].

As the gathered gods rejoiced at hearing Marduk's project to establish Babylon, he went on to assign them their duties:

Marduk, the King of the gods,
to Above and Below divided the Anunnaki.
To follow his instructions,
three hundred he assigned to the Skies,
as Those Who Watch he stationed them.

In like manner the stations on Earth he defined,
Six hundred of them on Earth he settled.

He issued all the instructions;
To the Anunnaki of Heaven and of Earth
he allotted their tasks.

The gods assigned to 'Mission Earth' are thus divided right off into two groups: Three hundred, named ***Igi.gi*** ("Those Who Observe and See'), have 'sky duties' and will be stationed "above the Earth" (on Mars, as we explain later). Six Hundred, the ***Anunnaki*** 'of Heaven *and* Earth', will be stationed on Earth itself; and their first task per their Lord's instructions, is to establish Babylon, and raise therein Marduk's stage-tower ***E.sag.il***—the 'House Whose Head Is Lofty'. (For depictions of Anunnaki and Igigi in their stations, see Fig. 64.)

By the end of Tablet VI *Bab-ili* (Babylon), the "Gateway of the gods," with its "Tower that reaches heaven," are ready; the Celestial Marduk is now also Marduk on Earth; and the reciting of *Enuma elish* proceeds to Tablet VII, which is a laudatory list of Fifty Names, fifty epithets of empowerment.

"With the title 'Fifty' the great gods proclaimed him (Marduk) supreme," the epic states in conclusion.

* * *

Obviously, the epic's Babylonian text has rushed events here 'fast forward'. Life has yet to emerge and evolve on Earth; Enki and his first crew of fifty Anunnaki are yet to splash down; cities of the gods need to be established; Man has yet to appear; and the Deluge still has to sweep over—for only in its aftermath does the episode of the Tower of Babylon take place. Whether the omissions are deliberate or not, the fact remains that all the interim developments still need to take place—not only according to the Bible, but also according to varied cuneiform texts.

Indeed, even before one contemplates the events on Earth, one ought to parse the enigma of events on Nibiru, where the coronation of Marduk presumably took place. Who are the assembled gods? Who are the 'Forefathers' that Marduk invoked? The divine-royal abode he plans to establish on Earth is to serve as a counterpart to the divine-royal abode of the god Anu, the ***E.sharra,*** on Nibiru. A king of what kingdom was Anu? Who were the Anunnaki and the Igigi, assigned to duties for Mission Earth? How did they come to be present, to reside, on planet Nibiru? Why did fifty of them—accompanying Ea/Enki—go to Earth in search of gold? And why, at its peak, were 600 Anunnaki and 300 Igigi needed?

While *Enuma elish* provides no such answers, we are not entirely at a loss to know them. Varied ancient texts fill-in data and details, name names and describe events. We have already mentioned some of those texts; we will bring to light many others—some even in languages other than Sumerian or Akkadian. Together they provide the dots that can be connected to form a coherent and continuous tale. Paramount in that context is what they tell us about ourselves—how Man and Mankind came to be on this planet Earth.

We can start unraveling the ball of yarn with Anu, the ruler on Nibiru during Marduk's confirmation as supreme leader of the Anunnaki and the Igigi. He was also ruler on Nibiru during the first arrival on Earth, for Ea/Enki invoked his status as "firstborn son of Anu" in his autobiography. One can assume that it was Anu's form of Kingship that was "brought down from heaven" by the Anunnaki, and it was from his court that the traditional insignia of kingship emanated: A divine headdress (crown, tiara); a scepter or staff (symbol of power, authority); and a coiled measuring cord (representing Justice); these symbols appear in divine investiture depictions at all times, in which the god or the goddess grants these objects to the new king (Fig. 56).

AN/Anu as a word meant 'Heaven'; as a name-epithet it meant 'The Heavenly One'; and its pictogram was a star. References in varied texts provide some information about Anu's palace, his court, and its strict procedures. We thus learn that in addition to his official consort (his spouse, Antu) Anu had six concubines; his offspring were eighty in number (only fourteen of whom bore the *divine titles* ***En*** for males or ***Nin*** for females [Fig. 57]). His court aides included a Chief Chamberlain, three Commanders in charge of the Rocketships, two Commanders of the Weapons, a 'Minister of the Purse' (= Treasurer), two Chief

Figure 56

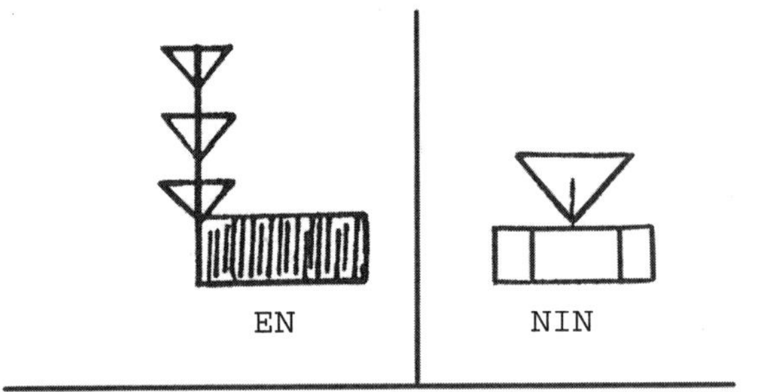

Figure 57

Justices, two 'Masters of Written Knowledge', two Chief Scribes, and five Assistant Scribes. The rank and file of Anu's staff were termed ***Anunna***—meaning 'Anu's Heavenly Ones'.

Anu's palace was located in the "Pure Place." Its entrance was constantly guarded by two royal princes; titled "Commanders of the Weapons," they controlled two divine weapons, the ***Shar.ur*** (= 'Royal Hunter') and the ***Shar.gaz*** (= 'Royal Smiter'). An Assyrian drawing (Fig. 58), purporting to depict the gateway to Anu's palace, showed its two towers flanked by "Eaglemen" (= uniformed Anunnaki 'astronauts'), with the Winged Disc emblem of Nibiru centrally displayed. Other celestial symbols—a twelve-member solar system, a crescent (for the Moon) and seven dots (for Earth) complete the presentation.

When an Assembly of the gods was called, it took place in the Throne Room of the palace. Anu sat on his throne, flanked by his son Enlil seated on the right and his son Ea seated on his left. Texts that recorded Assembly proceedings indicate that virtually anyone present could speak up; some of the deliberations were heated debates. But in the end Anu's word was final—"his decision was binding." Among his epithets was "Divine 60"—granting Anu, under the sexagesimal ('Base 60') numbering system, the highest rank.

The Sumerians and their successors have kept not only meticulous King Lists; they also maintained elaborate *God Lists*—lists of gods arranged by importance and rank and grouped by families. In the more detailed lists, the prime name of the god or goddess was followed by their epithets (that could be quite numerous); in some lists that attained a canonical status, the gods were arranged genealogically—giving, so to say, their royal pedigree.

There were local god lists and national god lists, short ones and

Figure 58

long ones. The most comprehensive, known to scholars by its opening line as the series *An:god-Anu* and deemed the Great God List, occupies seven tablets and contains more than 2,100 names or epithets of gods and goddesses—a mind-boggling number for sure, but considerably misleading if one realizes that sometimes a score or more listings were really epithets for the same deity (the younger son of Enlil, for example, who was called **Ishkur** in Sumerian, ***Adad*** in Akkadian, and *Teshub* by the Hittites, had another 38 epithet-names). The Great God List also included the deities' spouses and offspring, chief 'viziers' and other personal attendants.

Each tablet of this series is divided into two vertical columns, the one on the left giving the deities' Sumerian name/epithet and the matching one on the right the equivalent name or meaning of the epithet in Akkadian. Among other shorter or partial god lists thus far discovered was also one known as the series *An:Anu Sha Ameli;* despite its Akkadian title, it is an earlier basic listing of the Sumerian pantheon (listing only 157 names and epithets).

It is from such lists that we learn that the names chosen in *Enuma elish* for various planets were not accidental; they were names borrowed from the canonical god lists in order to enhance the genealogical claims of Marduk to supremacy—his being the son of Ea/Enki, in turn the firstborn son of Anu, who in turn was the scion of a royal Nibiruan line *of twenty-one predecessors*!

The list (arranged by couples) includes besides Anshar and Kishar, Laẖma and Laẖama (familiar as celestial names from *Enuma elish*) also unfamiliar names such as An.shargal and Ki.shargal, En.uru.ulla and Nin.uru.ulla; and (significantly) a couple oddly named **Alala** and **Belili**. This list of Anu's predecessors ends with the postscript "*21 en ama aa*"—***'twenty-one lordly mothers and fathers'*** (arranged as ten couples plus an unespoused male one). The Great God List then names the children and functionaries of Anu's Group, skipping his two principal sons and daughter (**Ea/Enki, Enlil** and **Ninmaẖ**), who are listed separately with their own family groups and aides.

Whichever way these god lists are studied, the major and dominant position of the divine king Anu is unmistakable. Yet a text titled *Kingship in Heaven,* found intact in a Hittite version, reveals that Anu was a usurper, having seized the throne on Nibiru by forcefully deposing the reigning king!

After calling upon the "twelve mighty olden gods," "the god fathers and the god mothers," and "all the gods who are in heaven and those upon the dark-hued Earth," to pay heed to the account of the usurpation, the text went on to say:

> Formerly, in the olden days,
> Alalu was king in heaven.
> Alalu was seated on the throne.
> Mighty Anu, first among the gods,
> stood before him, bowed at his feet,
> set the drinking cup in his hand.
>
> For nine counted periods
> Alalu was king in heaven.
> In the ninth period, Anu gave battle to Alalu.
> Alalu was defeated, he fled before Anu.
> He descended to the dark-hued Earth—
> down to the dark-hued Earth he went.
> Upon the throne Anu was seated.

Serving, then, as the royal Cup Bearer—a task calling for utmost loyalty—Anu betrays the king's trust and seizes the throne in a bloody *coup d'etat.* Why? Though he bears the epithet-title "First among the gods," the text fails to reveal the relationship between Anu and the reigning king; but the narrator's appeal to the Olden Gods, "The fathers and the mothers" of the gods, indicates a conflict or struggle over the throne whose roots go back several generations—a conflict caused by past events, genealogical relationships, or dynastic rivalries. With succession rules that tried to untangle conflicting claims between a firstborn and a legal heir, between a son by a spouse and another by a concubine, and a rule granting primacy to a son by a half-sister, Anu evidently had a claim on the throne that (in his view) trumped that of Alalu.

Such conflicting claims, one must conclude, began long before the Anu/Alalu incident and, as we shall see, continued after that. Certain aspects of the god lists serve as clues to an old and festering problem regarding Kingship on Nibiru—issues that in time affected events on Earth. In the Great God List (the extant version was probably compiled in Babylon) the Enki Group follows that of Anu's; the Enlilites come next, followed by Ninḫarsag's group. But in other lists—including the shorter Sumerian one—it is the Enlil Group that follows Anu's. These varied positionings reflect a tug-of-war that calls for a closer look.

The Great God List contains another a puzzling feature: When it comes to Enki (but not so for Enlil) it inserts into his listings the names of predecessor ancestor-couples that are different from those of Anu's; they bear such names as En.ul and Nin.ul, En.mul and Nin.mul, En.lu and Nin.lu, En.du and Nin.du, etc. These are divine predecessor couples of Enki that are not found in the Anu group. It is only when the list comes to the tenth couple, named **Enshar** and **Kishar,** that an apparent match with **Anshar** and **Kishar** in Anu's list occurs. Since Anu was Enki's father, the separate or non-Anu ancestor couples had to represent the line of Enki's mother, who had to be someone other than Antu—in other words, a concubine. That, it became clear as events unfolded, was a serious hierarchial defect.

In his autobiography Enki declared, with some desperation: "I am the leader of the Anunnaki, engendered by fecund seed—*the firstborn son of divine An,* the Big Brother of all the gods." Firstborn indeed he was; engendered by "fecund seed" he was—but only from his father's side. When it came to be seated beside the enthroned Anu, it was Enlil who sat on the right. In the numerical ranking of the elite Twelve Great Gods, Enlil was second to Anu with the rank of 50; Enki followed with the lesser rank of 40. Though Enki was the firstborn, he was not the Crown Prince; that title with the right of succession was granted to the younger Enlil because his mother was **Antu**—and Antu was not just Anu's official spouse, *she was also a half-sister of Anu,* providing Enlil with a double dose of the "fecund" genetic seeds.

A picture thus emerges of two old-time clans, vying for Kingship on Nibiru; at times at war, at times seeking peace through intermarriage (a device not unknown on Earth, where warring tribes or nations often resorted to royal intermarriage to bring peace), and taking turns on the throne—sometimes violently, as in the case of Anu's coup against Alalu. The name of the deposed king (**Alalu** in Hittite) is clearly different from the many '*En-*' ones, but is virtually identical to the oddly named **Alala** in Anu's list, suggesting affiliation to a different clan and access to the throne through intermarriage.

That emphasis on one's genetic "seed" line and Succession Rules was reflected in the Bible's tales of the Patriarchs.

* * *

Was the violent overthrow of Alalu, causing him to flee his home planet, an isolated event—or an episode in a history of continuous (even if intermittent) fighting between two clans, perhaps—in planetwide terms—between two nations on Nibiru? The data in the God Lists suggests that his overthrow was a continuation of unresolved strife between the Niburian clans. It was neither the first nor the last violent 'regime change': Some texts suggest that Alalu himself was a usurper, and that later on attempts were made to overthrow Anu . . .

A detail in the makeup of Anu's royal court offers a clue to events on Nibiru: It is the listing of three "Commanders in charge of the ***Mu*** rocketships" and two "Commanders of the Weapons." Come to think of it, it means that five military men made up almost half the ministerial cabinet of eleven (we exclude the seven scribes). This is tantamount to a military government. There is an obvious stress on weaponry: Two of the five generals deal just with weaponry. When it comes to the palace proper, it was protected by two awesome weapons systems, overseen by two royal princes.

Protected from what? Protected from whom?

At the risk of preempting a chapter yet to follow, we can mention already here that in 2024 B.C. the Anunnaki then on Earth resorted to the use of nuclear weapons in their continuing clan clashes. Several ancient texts (which we shall quote) state that seven nuclear devices were used; and it is clear that they were brought over to Earth from Nibiru. Whether or not the ***Sharur*** and ***Shargaz*** that protected Anu's palace were such weapons, it is evident that *nuclear weapons were part of the Nibiruan military arsenal.* Were they ever used on Nibiru? Why not, if they were used on a distant planet called Earth, on which at their peak just 900 Nibiruans (600 Anunnaki, 300 Igigi) were stationed? So much more was at stake on Nibiru itself!

From viewing our Solar System as a once-created/forever-frozen assemblage of planets orbiting a central nuclear cauldron (the Sun), space-age astronomers now realize that the planets and even their moons are alive with natural phenomena—have their own inner nuclear cores, create and emanate heat, sustain volcanic activity, have atmospheres, have changing climates; some display frozen surfaces, some display Earth-like features; many have water, some only chemical-filled lakes; some seem bone dead, some reveal complex compounds that could be associated with Life. Seasons have even been detected on 'Exoplanets' orbiting other distant star-suns—planets whose mere notion of their possible existence was the domain of science-fiction until a few years ago.

Our neighbor Mars, considered just decades ago a lifeless planet since its birth, is now known (thanks to unmanned space exploration since the 1970s) to have had a proper atmosphere (still sufficient to have occasional dust storms), flowing water, rivers, and vast seas and lakes—with a frozen lake, water ice, and even muddy soil to this very day (Fig. 59, sample scientific reports). It is noteworthy that in *The 12th Planet* (1976) we had already provided evidence that a habitable Mars served the Anunnaki as a way station for the interplanetary spacecraft from and to Nibiru; it was there that the Igigi were stationed, their task to operate smaller shuttlecraft between Earth and Mars.

On Earth, the Igigi landed their shuttlecraft on a vast platform with launch tower called 'The Landing Place', built of colossal stone blocks; we have identified it in *The Stairway to Heaven* as the site known as Baalbek in the Lebanon mountains. The vast stone platform still exists; so do the remains of the launch tower—built of immense stone blocks

Signs of Ancient Rain May Stretch Mars's Balmy Past

In the evolving debate over water on Mars, the Hesperian epoch of middle martian histo- show up in images from the Thermal Emission Imaging System (THEMIS) which ha- were cut. And the northern end of one

When water gushed on Mars

Were the northern plains of Mars submerged in a vast flood as recently as 20,000 years ago? Geologists claim to have found evidence of a recent volcanic eruption under the ice cap that could have cre- high and 35 kil- Signs of volc- water have be- study links the logical evidenc- was the chief s- the northern n- researchers sa- past 10 million 20,000 years ag- has not been a- and warm peri- Conditions flood — sulphu water, an ice ca- light —

ES, TUESDAY, NOVEMBER 4, 2008

re Recent Presence of Water

ONAL FRIDAY, AUGUST 1, 2008

Test of Mars Soil Sample Confirms Presence of Ice

By KENNETH CHANG

Heated to 32 degrees Fahrenheit, a sample of soil being analyzed by NASA's Phoenix Mars lander let out a puff of vapor, providing final confirmation that the lander is sitting over a large chunk of ice.

sure and wind speeds. The lander has had some difficulty shaking the clumpier-than-expected soil out of its scoop into the instruments for analysis. One sample of soil

Mars spacecraft traces a watery tale

A Mars-orbiting spacecraft is providing new details about when and where liquid water, an essentia- the planet detected a water-bea- have prov- Red Plan- Over t- rovers ha- that liqui- on Mars channel surface mark of altered

Mars lander confirms water ice

Phoenix also samples unexpected chemical compound

By Ashley Yeager

The Phoenix Mars Lander has finally confirmed the presence of water ice on scientists have

tasted ice on Mars," said Boynton, a Phoenix coinvestigator and the lead TEGA scientist. "And I can say it tastes very fine." One of the lander's instruments has also "tasted" an unexpected chemical com- The com-

Scientists: Fresh signs of water on Red Planet

John Johnson Jr.
LOS ANGELES TIMES

NASA scientists announced Wednesday that they had found evidence that water still flows over the surface of Mars — sporadic gushers th-

A Wetter, Younger Mars Emerging

rs seemed a liv- rs thought they huge areas each rough planet- caught everyone's attention. Other, perhaps more persuasive, signs also suggest that water may even now flow on or beneath the frigid surface. seeing more and more evidence splashing on the Rains may have lion years ago unde- mosphere: an ocea- northern lowlands a and water may have

PLANETARY SCIENCE

Martian Lake View

Mars seems to have a frozen lake on its surface, according to images obtained by the European Space Agency's Mars Ex-

Evidence for Recent Groundwater Seepage and Surface Runoff on Mars

Relatively young landforms on Mars by the Mars Global Surveyor the presence of n in high-resolution images acquired Camera since March 1999, suggest shallow depths beneath titudes (particularly a very small numl ger martian valleys ocesses associated ive youth of the otherwise geolog- ndforms or cross- and eolian dunes. argue for constrai-

An Early, Muddy Mars Just Right for Life

When the Opportunity rover found the salty sedimentary remains of standing water on Mars, the prospects for early life on another planet brightened considerably. Although mud-laden, those early waters were nothing have adapted to.

First hard evidence found of a lake on Mars

Wed Jun 17, 2009 7:29pm EDT

WASHINGTON (Reuters) - A long, deep canyon and the remains of beaches are perhaps the clearest evidence yet of a standing lake on surface of Mars -- one that apparently contained water when the pl-

Figure 59

that range from 600 to 900 tons each. At the norhwestern corner of the platform, the tower was reinforced with three gigantic stone blocks weighing more than 1,100 tons each (!); known as the Trilithon (Fig. 60), local lore attributes them to "the giants."

Our own planet, Earth, has undergone a violent beginning, the gathering of oceans and seas, the rise and shifting of continents ("firm

Figure 60

land"), volcanic eruptions and tidal waves (remember the Deluge?), Ice Ages and warm intervals (alias Climate Change), and atmospheric problems due to too much of this (e.g., carbon emissions) or too little of that (such as loss of protective ozone). It is only logical to assume that planet Nibiru underwent similar natural events.

Some who have read *The 12th Planet* and accepted its conclusions regarding Nibiru still wondered how the Anunna could survive on a planet whose orbit takes it far away from the Sun; wouldn't they, and all life, freeze to death right off? My answer has been that we and life on Earth face the same issue even though Earth is at a presumed "livable distance" from the Sun; all we have to do is leave Earth's surface a little bit, and we'll freeze to death. Earth, like other planets, has a nuclear core that produces heat—it gets warmer and warmer as miners tunnel deeper down. But our very thick rocky mantle makes us dependent on heat coming from the Sun. *What protects us is Earth's atmosphere:* Acting as a greenhouse, it keeps in the warmth we get from the Sun.

In the case of Nibiru, it is again the atmosphere that offers protection; but there, the need is mostly to keep in the heat coming from the planet's core and prevent it from dissipating out into space. For it is only for part of its 'year' (one orbit around the Sun) that Nibiru's elliptical orbit (see Fig. 52) provides a warm 'summer'; during its much longer 'winter', the planet depends on its inner-core heat to keep its life going.

As all planets, Nibiru too must have undergone natural climate and atmospheric changes; when its inhabitants became capable of manned space flight and attained nuclear technology, the use of nuclear weapons made atmospheric problems worse. It was then, I suggested in *The 12th Planet,* that Nibiru's scientists came up with the idea of creating a shield of gold particles to mend and protect their planet's damaged atmosphere. But gold was a rare metal on Nibiru, and its use or misuse for the planet's salvation only added to the simmering conflicts.

It was against such a background of circumstances and events that Anu seized the throne from Alalu; and Alalu, escaping for his life in a rocketship, sought haven on a distant and uninhabited strange planet. The Nibiruans called the distant planet ***Ki;*** the ancient Hittite text made clear that "down to the dark-hued Earth Alalu went." His chance discovery that its waters contained gold served as a trump card for demanding reinstatement to Kingship. In *The Lost Book of Enki* I have suggested that Alalu agreed to let Ea come to verify the discovery because was Ea was his son-in-law, having espoused—for state reasons—Alalu's daughter **Damkina**. In the post-overthrow circumstances of mistrust and animosity, Ea/Enki—a son of Anu, son-in-law of Alalu—was perhaps the only one trusted by both sides to lead Mission Earth. And so it was that Ea and his crew of fifty came to Earth to retrieve and send back to Nibiru the invaluable metal—a mission and an arrival described by Ea in his autobiography.

From then on, the main stage for the subsequent astounding events was Planet Earth.

* * *

As great a scientist as Ea was, he could not extract from the waters of what we now call the Persian Gulf more gold than it contained—minute quantities requiring the processing of huge volumes of water. A great scientist that he was, Ea traced the gold to its nearest prime source—the gold lodes deep in the rocks of the Abzu. If Nibiru must have the gold—as it surely did—the Anunnaki had to switch to a mining operation and establish ***Arali,*** the Land of Mines.

The changed nature of Mission Earth required more personnel, new equipment, settlements on two continents, new transportation and communication facilities; it all required a different type of leader—one less of a scientist and more with organizational, discipline, and command experience. The one chosen for the task was **En.lil** (= 'Lord of the Command'), the Crown Prince. Subsequent events showed him to be a strict disciplinarian, a 'by-the-book' commander.

While Enki's coming to Earth is documented in his inscribed autobiography, Enlil's journey is recorded in another kind of document. It is an unusual circular tablet, a disc made of an unusual kind of clay. Found in the ruins of Nineveh (sketch, Fig. 61) its present keeper, the British Museum in London, displays it just as a sample of ancient writing—an incredible act of missing the point, for ***the artifact provides a unique depiction of the heavens in which the route of Enlil from his planet to Earth is described both graphically and in words!***

It is divided into eight segments; the information regarding Enlil's journey is found in a segment that fortunately is mostly undamaged. At the segment's margins stars and constellations are named, indicating that the celestial space is out there. The writings on the sides (in translation, Fig. 62) suggest landing instructions. In the segment's center a route is drawn connecting the pictograph for "moutainous planet" to a segment of the skies familiar from Sumerian astronomy as Earth's location. The route's course takes a turn between two planets whose Sumerian names stand for Jupiter and Mars. And the statement (in Akkadian) under the route line clearly says: ***"The god Enlil went by the planets."*** There are seven of them—accurately counted, since for anyone coming into our Solar System from its outer range, Pluto would be the first planet

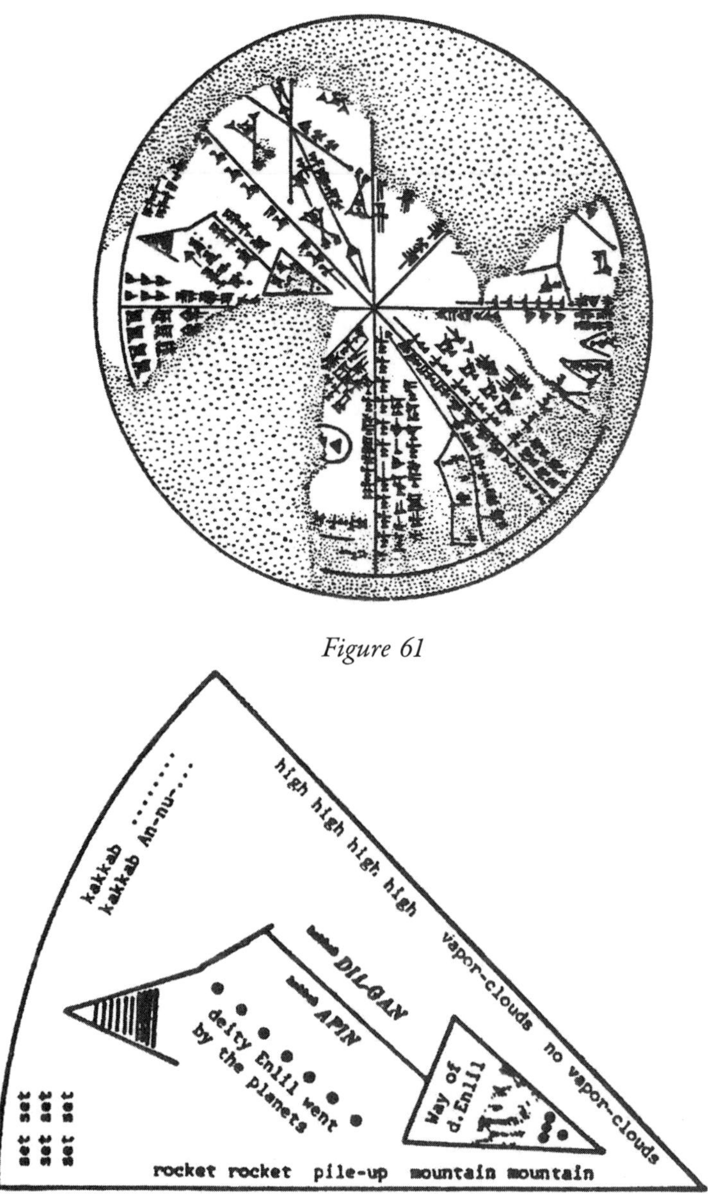

Figure 61

Figure 62

encountered, Neptune and Uranus second and third, Saturn and Jupiter fourth and fifth, Mars the sixth, and Earth the seventh.

* * *

The change in duties and command structure was, at best, not an easy undertaking. It was doubly difficult to diminish Ea's prerogatives by sending to Earth his rival for the crown—Enlil. The bickering and mistrust between the half-brothers is reflected on the one hand by Enki's cry that he the firstborn, "fecund seed," is now reduced in status; and by Enlil, in a text recording his complaint that Ea is witholding from him the ***Mé***—an enigmatic term usually translated 'Divine Formulas'—some kind of 'memory chips' essential for every aspect of the mission. Matters got so bad that Anu himself journeyed to Earth and offered his two sons to ***settle the issue of succession by drawing lots***. We know that, and we know essentially what ensued, from the *Atra-Hasis Epic:*

> The gods clasped hands together,
> cast lots and then divided:
> Anu, their father, was the king;
> Enlil, the Warrior, was the Commander.
>
> Anu went up [back] to Heaven,
> the Earth [he left] to his underlings.
> The seas, enclosed as with a loop,
> To Enki the prince were given.
> After Anu had gone up to Heaven,
> Enki to the Abzu went down.

The text's subsequent fourteen lines, that certainly dealt with Enlil's domain and tasks, are too damaged to be fully read and translated. But the legible portions of other lines indicate that while Ea—renamed Enki (= 'Lord [of] Earth') as a solace—was assigned to the Abzu to oversee the mining operation, Enlil took charge of the *Edin,* whose two rivers, the Euphrates and Tigris, are clearly mentioned. We know from other texts that Enlil increased the number of Anunnaki settlements there from Ea's sole Eridu to the famed five Cities of the Gods, and then added three more—Larsa, Nippur, and Lagash.

Nippur (Akkadian from the Sumerian ***Ni.ibru*** = 'The Splendid

Place of Crossing') served as Enlil's Mission Control Center. The Anunnaki built there the ***E.kur*** (= 'House which is like a mountain'), a temple-tower whose "head was raised" heavenward; its innermost chamber, equipped with 'Tablets of Destinies' and humming with other instruments emitting a bluish light, served as the ***Dur.an.ki***—the 'Bond Heaven-Earth'. Having been forced to provide Enlil with the essential ***Me,*** Enki (his autobiography states) "filled the Ekur, abode of Enlil, with possessions"; and the "boats of Meluhha, transporting gold and silver, brought them to Nippur for Enlil."

When the eight settlements are pinpointed on a map, a purposeful layout emerges (Fig. 63). Nippur was physically at the center; the others, located in concentric distances, formed a flight corridor; leading to

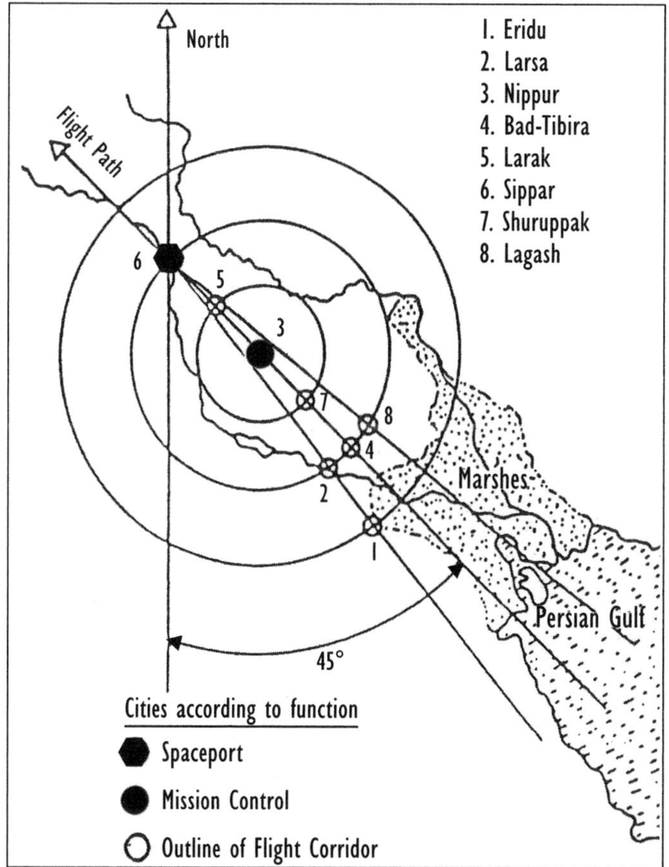

Figure 63

Sippar (the Spaceport-city), it was anchored on the peaks of Mount Ararat (highest topographical feature in the Near East). Medical facilities were at Shuruppak. Bad-Tibira was the metallurgical center where ores from the Abzu were processed; from Sippar, the ingots were regularly transported in small shipments to Mars—for Mars, with its lesser gravitational pull, served as a space base from which the Anunnaki shipped larger and heavier gold loads to Nibiru.

Arriving in groups of fifty, the *Anunna* were divided into two groups. Six hundred, henceforth known as the ***Anunnaki*** (= 'Those who from Heaven to Earth came'), were based and served on Earth; their assignments included mine work in the Abzu and the tasks in the Edin. Another three hundred, desigated ***Igi.gi*** (= 'Those Who Observe and See') operated the shuttlecraft between Earth and Mars—and their main base was on Mars.

The setup is depicted on a 4,500-year-old cylinder seal, now kept at the Hermitage Museum in St. Petersburg, Russia (Fig. 64). It shows an Anunnaki 'Eagleman' (astronaut) on Earth (symbolized by seven dots and the Moon's crescent) greeting a mask-wearing Igigi 'Fishman' on Mars (the six-pointed planet symbol); a circular spacecraft with extended panels is shown in the skies between them.

As Mission Earth was in full swing, Nibiru was saved; but on Earth itself, trouble was brewing.

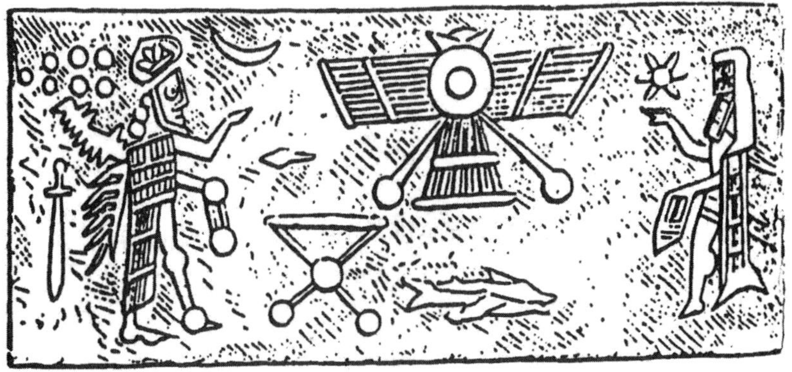

Figure 64

THE TALE OF THE EVIL ZU

A Sumerian text known as *The Myth of Zu* is a source of information about Enlil's ***Duranki*** as well as about the ***Igigi*** and the weapons of the ***Anunnaki***. It deals with an attempted coup against Enlil by an Igigi leader named **Zu**. (A recent discovery of the text's tablets suggests that his epithet was **An.zu** = 'Knower of Heaven'.)

Based on Mars where they had to wear spacesuits with breathing masks (see Fig. 64), and confined on Earth to the 'Landing Place' in the cedar mountain, "the Igigi, one and all, were upset"—they were complaining and restive. Their leader, Zu, was invited to Enlil's headquarters to talk things over. Trusted enough to freely pass through the guarded entrance, the "evil Zu to remove the Enlilship"—to seize the command—"conceived in his heart: To take the divine Tablet of Destinies, to rule the decrees of all the gods . . . to command all the Igigi."

And so, one day when Enlil was bathing, "Zu seized the Tablet of Destinies in his hands, took away the Enlilship," and flew away with it to the hideaway in the mountains. The removal of the Tablet of Destinies caused a flash of "blinding brightness" and brought the *Duranki* to a standstill:

> Suspended were the Divine Formulas;
> The sanctuary's radiance was taken off;
> Stillness spread all over; silence prevailed.

"Enlil was speechless. The gods of the land gathered at the news." Alarmed by the gravity of the usurpation, Anu sought a volunteer among the gods to challenge Zu and retrieve the Tablet of Destinies; but all who tried failed, for the Tablet's mysterious powers warded off all projectiles shot at Zu. Finally, Ninurta, Enlil's firstborn, using his "seven-cyclones weapon" (see illustration), created a dust storm that forced Zu to take flight "like a bird." Ninurta pursued him in his skyship, and an aerial battle ensued. Shouting "wing to wing!" Ninurta shot a *Til.lum* (= 'Missile') at Zu's "pinions," causing Zu to crash to the ground. He was captured by Ninurta, tried, and sentenced to death. The Tablet of Destinies was reinstalled in the Duranki.

Echoing the Sumerian Tale of Zu, the lores of other peoples also relate divine aerial duels. The Egyptian hieroglyphic text 'The Contending of Horus and Seth' describes the defeat of Seth by Horus in an aerial battle over the Sinai peninsula. In Greek tales of the gods, the fierce battles between Zeus and the monstrous Typhon ended when Zeus, in his Winged Chariot, shot a thunderbolt at the magical aerial contraption of his adversary. Aerial battles between gods flying in "cloud-borne chariots" and using missiles are also described in the Hindu Sanskrit texts.

VIII
A Serf Made to Order

The unrest among the Igigi that led to the Zu Incident was only a prelude to other troubles involving them—troubles inherent in long-term interplanetary missions; and the absence of female companionship turned out to be one of the major issues.

The problem was less acute in the case of the Earth-stationed Anunnaki, for there were females among them from the very first landing party (some of whom are mentioned by name and tasks in Enki's autobiography). Additionally, a group of female nurses, led by a daughter of Anu, were sent to Earth (Fig. 65). Her name was **Ninmah** (= 'Mighty Lady'); her task on Earth was that of ***Sud*** (= 'One who gives succor'): She served as the Anunnaki's Chief Medical Officer and was destined to play a major role in many of the subsequent events.

But trouble brewed also among the Earth-based Anunnaki, especially

Figure 65

among those assigned to mining duties. The *Atra-Hasis Epic* in fact tells the story of a ***mutiny of the Anunnaki*** who refused to go on working in the gold mines and the ensuing chain of unintended consequences. The epic's ancient title echoed its opening words *Inuma ilu awilum* ('When the gods, like men'):

> When the gods, like men,
> bore the work and suffered the toil—
> The toil of the gods was great,
> the work was heavy, distress was much.

The irony in the title is that the gods toiled as though they were men because there were yet no men on Earth. The Epic's tale is in fact the tale of the Creation of Man to take over the gods' toil. Indeed, the very Akkadian term *Awilu* means 'Workman'—a toiler—rather than simply 'Man' as is usually translated. The feat that changed everything was an accomplishment of Enki and Ninmah; but as far as Enlil was concerned, it was not a tale with a happy ending.

As the Anunnaki miners "toiled deep in the mountains, they counted the periods of the toil." "For 10 periods they suffered the toil, for 20 periods they suffered the toil, for 30 periods they suffered the toil, for 40 periods they suffered the toil":

> Excessive was their toil for 40 periods,
> [. . .] they suffered the work night and day.
> They were complaining, backbiting.
> Grumbling in the excavations (they said):
> "Let us confront [. . .], the Commander,
> that he may relieve us of our heavy work.
> Let us break the yoke!"

The occasion for the mutiny was a visit by Enlil to the mining area. "Come, let us unnerve him in his dwelling!" a ringleader (whose name is illegible in the tablet) urged the angry miners. "Let us proclaim a mutiny, let us adopt hostilities and battle!"

The gods heeded his words.
They set fire to their tools,
put flame to their earthcutters
and fire to their grinders.
Throwing them away, they went
to the gate of the hero Enlil.

It was nighttime. As the mutineers reached the place where Enlil was staying, the gatekeeper Kalkal barred the gate and alerted Enlil's aide **Nusku**, who awakened his master. Hearing the shouting—which included calls to "kill Enlil!"—Enlil was incredulous: "Is it against me that it is being done? What do my own eyes see?" Through Nusku he demanded to know "Who is the instigator of this conflict?" The mutineers responded by shouting, "Every single one of us has declared battle . . . Our work is heavy, distress is great—excessive toil is killing us!"

"When Enlil heard those words, his tears flowed." Contacting Anu, he offered to resign his command and return to Nibiru, but demanded that the instigator of the mutiny be "done to death." Anu summoned the Council of State. They found that the Anunnaki's complaints were justified; but how could the vital gold-supply mission be abandoned?

It was then that "Enki opened his mouth and addressed the gods his brethren." There is a way out of the dilemma, he said. We have with us Ninmah̲; she is *Belet-ili,* 'a Birth-Giving goddess'—

Let her fashion a *Lulu,*
Let an *Amelu* bear the toil of the gods!
Let her create a *Lulu Amelu,*
Let him bear the yoke!

He was suggesting to create a *Lulu*—a "Mixed One," a hybrid—to be an *Amelu,* a workman, to take over the Anunnaki's toil.

And when the other gods asked how such a *Lulu Amelu* could be created, Enki answered: "**The creature whose name you uttered—it exists!**" All we need to do is "**bind on it the image of the gods.**"

Therein, in this response, lies the answer to the enigma of 'The

Missing Link'—how could *Homo sapiens,* modern man, appear in southeast Africa some 300,000 years ago ***overnight*** (in anthropological terms) when the evolutionary advances from apes to hominids, and in hominid species from *Australopithecus* to *Homo habilis* to *Homo erectus,* etc., took millions upon millions of years?

A Being, akin to the Anunnaki in many respects, Enki told the astounded gods, *already exists* in the wilds of the Abzu. "All we need to do is ***bind on it the image the gods***"—to upgrade it with some Anunnaki genes—and create a ***Lulu*** (= 'A Mixed One') who could take over the mining work.

What Enki had discovered at his headquarters in southeast Africa was a hominid so akin genetically to the Anunnaki, that with some genetic tinkering—adding to the genome of the hominid (say a *Homo erectus*) some Anunnaki genes—could upgrade the hominid to the status of an understanding, speaking, tool-handling *Homo sapiens.* And it was all possible because the DNA on Earth was that of Nibiru, transferred—the reader will recall—when Nibiru itself smashed into Tiamat!

Enki then outlined to the assembled leaders how it could be done with the help of Ninmah and her biomedical expertise. Hearing that,

> ***In the Assembly,***
> ***the Great Anunnaki***
> ***who administer destinies,***
> ***declared: "YES!"***

That fatal decision to create Man is echoed in the Bible. Identifying the assembled Great Anunnaki as the *Elohim,* the 'Lofty Ones', Genesis 1:26 states:

> And *Elohim* said:
> "Let us make an *Adam*
> in our image
> and after our likeness."

There is no doubting the plural in the biblical statement, starting with the plural ***Elohim*** (the singular is *El, Elo'ha*) through "Let ***us*** make"—"in ***our*** image"—"and ***our*** likeness." It happened "40 periods"—40 *Shars*—after the arrival of the Anunnaki. If the Arrival (see previous chapters) took place some 445,000 years ago, the creation of *Adamu* took place 300,000 years ago (445,000—144,000)—exactly when *Homo erectus* suddenly changed to *Homo sapiens.*

* * *

The process by which the fashioning of the "Primitive Worker" was achieved is then described in the Atra-H̲asis Epic, as well as in several other texts. It involved obtaining from the blood of a god his *Te'ema*—a term scholars translate as 'Personality' or 'Life's Essence'—and mixing it with the "*Ti-it* of the Abzu." The term *Ti-it* has been presumed to come from the Akkadian word *Tit* = clay, hence the notion (echoed in the Bible) that 'The Adam' was fashioned from clay or 'dust' of the Earth. But read in its Sumerian origin, ***Ti-it*** means "That which is with life"—the 'essence' of a living being.

The *Te'ema*—the 'Life's Essence' or 'Personality' of a god—what we would now define as ***his genetic DNA***—was "mixed" with the 'essence' of an existing Being found (the text states) in the area "just above the Abzu." ***By mixing genes extracted from the blood of a god with the 'essence' of an existing earthly being, 'The Adam' was genetically engineered.***

There was no 'Missing Link' in our jump from *Homo erectus* to *Homo sapiens,* because the Anunnaki jumped the gun on Evolution through genetic engineering.

The task described by Enki was easier said than done. In addition to the Atra-H̲asis epic, other texts detail the creation process. Extensively rendered in both *The 12th Planet* and *Genesis Revisited,* they describe considerable trial and error, resulting in beings missing limbs, with defective or odd organs, or with flawed eyesight or other senses. As the experiments continued, Ninmah̲ figured out which genes affect what, and declared that she now could deliberately produce—"as my heart desires"—beings with or without this or that defect . . .

Enki, a text states, "prepared a purifying bath" into which "one god was bled." Ninmaẖ "mixed blood and flesh" in order to "fix upon the newborn the image of the gods." Enki "was seated before her; he was prompting her" with instructions and advice. The genetic endeavor was conducted in *Bit Shimti,* a laboratory-like place whose Sumerian name ***Shi.im.ti*** literally meant "Place where the Wind of Life is blown in"—a detail from which the biblical verse about "blowing the Breath of Life" into The Adam's nostrils (Genesis 2:7) was in all probability taken.

Ninmaẖ was handling the mixing; "reciting the incantations," Ninmaẖ was listening for an *Uppu*—a heartbeat. When the "Perfect Model" was finally attained, Ninmaẖ lifted him and shouted, "I have created! My hands have made it!" (Fig. 66).

Announcing the feat to the great gods, here is what she said:

> You commanded me a task;
> I have completed it . . .
> I have removed your heavy work.
> I have imposed your toil on *Awilum* ('Work-man').
> You raised a cry for *Awiluti* ('Mankind')—
> I took off your yoke, I established your freedom!

"When the gods heard this speech of hers, they ran together and kissed her feet." They called her **Mami** (= 'The Mother'), and renamed her **Nin.ti** (= 'Lady of Life'). The solution suggested by Ea was achieved.

Figure 66

The genes we got were those of a male Anunnaki (lately discovered Atra-Ḫasis tablets reveal that he was the leader of the mutiny); *but with all due respect to a male God or god,* ***it was a female goddess who had actually created us.***

* * *

It required additional genetic engineering—even some surgery under anesthesia (reported both in a Sumerian text and in the Bible)—to fashion a female counterpart; but like hybrids to this day (such as a mule, the 'mixed' product of a horse and a donkey), they could not procreate. To make 'copies' of the Perfect Model of the *Lulu Amelu,* difficult and time-consuming reproduction by young "birth goddesses" was required. The next step of genetic engineering—enabling the *Lulus* to procreate on their own—was undertaken by Enki, the 'Serpent' in the biblical Garden of Eden version.

As the biblical tale has it, The Adam who was placed in the orchard of the gods to till it and to tend it, was warned by God (the Hebrew term is actually *Yahweh Elohim*) not to eat of the Tree of Knowing, "for on the day you eat thereof surely you shall die." Put into deep sleep, the Adam is operated upon, and a counterpart female is fashioned from his rib. The Adam and "the woman" (she is not yet named!) go about naked "and are not ashamed."

The wiley Serpent now approaches the woman regarding the prohibited tree, and she confirms that that is what *Elohim* had said. But "the Serpent said to the woman: No, you will not die!" So the woman, seeing that the Tree's fruit was edible, "took of its fruit, and ate, and also gave to her mate, and he ate." And right away they became aware of their sexuality; realizing that they were naked, they made themselves aprons out of fig leaves.

It was those aprons that gave them away; for the next time *Yahweh Elohim* saw them, he noticed that they were no longer naked; questioning The Adam about it, he found out what had happened. Angered, "What have you done!" God shouted at the woman—because of that, "in pain and suffering you will bear children." Alarmed, God said to unnamed colleagues: "Behold, The Adam has become as *one of us* to know good and evil; what if he put forth his hand and took also of the Tree of Life, and ate, and lived forever?" And God expelled The Adam and Eve from the Garden of Eden.

The tale, without doubt, explains how Adam and Eve were enabled to procreate—a development blamed, in the Bible on the 'Serpent,' the Hebrew word for which, *Nachash,* could also mean "He who solves puzzles." Not surprisingly, the Sumerian parallel for these varied meanings also comes from a single term—**Buzur**—which was an epithet of Enki meaning "He who solves secrets." The hieroglyph for ***Ptaẖ,*** his Egyptian name, was an Entwined Serpent. In the Mesopotamian texts, Enki was assisted in this secret knowledge by his son **Nin.gish.zidda** (= 'Princely Lord of the Tree of Life') whose emblem—Entwined Serpents—has remained the symbol of medicine to this day. Without doubt, these name meanings and Entwined-Serpent emblems are echoed in the biblical tale of the Serpent and the two special Trees in the Garden of Eden. And now that modern science has discovered the structure of DNA strands, it is possible to realize that Ningishzidda's emblem of two entwined serpents is in fact a rendering of the two-stranded, entwined *double-helix DNA*. We demonstrate their similarities in Fig. 67.

"Out of the god's blood they fashioned Mankind," the texts reiterate; "they imposed on it the tasks, to let free the gods; it was a work beyond

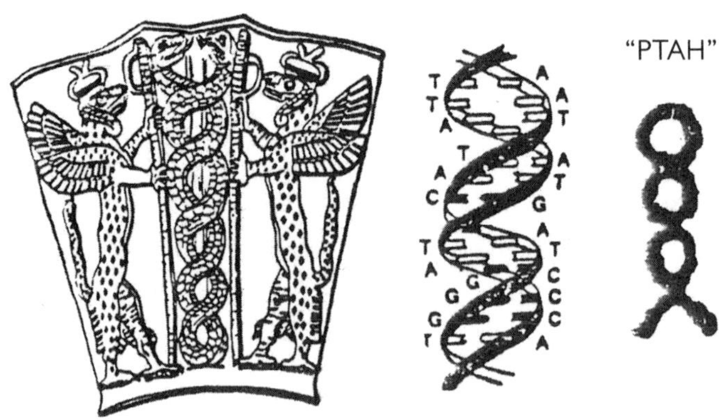

Figure 67

comprehension." Indeed it was; and it happened some 300,000 years ago—*just when* Homo sapiens *suddenly apeared in southeast Africa.* It was then that the Anunnaki 'jumped the gun' on Evolution and, using genetic engineering, upgraded a hominid—say *Homo erectus*—to an intelligent, tool-handling *Homo sapiens* (= 'Wise Man') to be their serf. It happened in the area "above the Abzu"—exactly where fossil remains indicate: In the Great Rift Valley zone of southeast Africa, just north of the gold-mining land.

* * *

We know from the continuing *Atra-Hasis* text and from other detailed texts that no time was lost in putting the Primitive Workers to work in the mines, and that Anunnaki from the settlements in the *Edin* raided the mines and forcefully brought some of those workers to serve them in the Edin, where "with picks and spades they built the shrines, they built the canal banks, they grew food for the people and for the sustenance of the gods."

The Bible, though more briefly, reports the same: "And Yahweh *Elohim* took The Adam"—from where he had been created—"and placed him in the garden of Eden, to till it and to tend it." (The Bible specifically precedes here 'Adam'—'He of Earth', an Earthling—with the definitive article '*The*', making clear it is a species that is written about, as distinct from a person named 'Adam', husband of Eve, whose tale starts only in chapter 4 of Genesis.)

"***To till it and to tend it,***" to be an *Amelu*, a workman. The Bible has similarly stated: "*Adam le **amal** yulad*"—'Adam to toil was created'. And the Hebrew term *Avod,* translated 'worship', in truth means 'To work'.

Man was fashioned by the gods to be their serf.

* * *

Time passed; "the [settled] land extended and the people multiplied." Thus does the *Atra-Hasis* epic start the next phase of the events that followed the Mutiny and the Creation of The Adam, and that finally led to the Deluge.

The people, in fact, multiplied so much (the text reports) that "the land was bellowing like a bull." Enlil was not happy: "the god was disturbed by their commotion." He made his displeasure known: "Enlil heard their bellowing and said to the great gods: 'The bellowing of Mankind has become too intense for me; by their commotion, I am deprived of sleep." Of the damaged lines that follow, only Enlil's words "let there be a plague" are legible; but we know from the parallel biblical narrative that "Yahweh repented that He had made The Adam on Earth . . . and said: I will wipe The Adam that I have created off the face of the Earth" (Genesis 6:6–7).

The tale of the Deluge and its hero (Noah/Utnapishtim/Ziusudra) is told in both sources along similar lines, except that unlike the monotheistic Bible where the same God first decides to destroy Mankind and then saves it through Noah, the Mesopotamian version clearly identifies Enlil as the angry deity—while it is Enki, defying Enlil, who saves the "Seed of Mankind." On the other hand, the biblical narrative (which compresses all the deities into a sole God) provides a more profound reason than 'bellowing' or 'commotion' for the dissatisfaction with Mankind. In the words of chapter 6 of Genesis, it came to pass that

> When The Adam began to multiply
> on the face of the Earth
> and daughters were born unto them,
> that the sons of the *Elohim*
> saw the daughters of men
> that they were suitable,
> and they took them as wives
> of all which they chose.

Yahweh, Genesis tells us, was angered by what was going on: "Yahweh saw that the Wickedness of Man was great upon the Earth . . . and Yahweh repented that He had made The Adam on Earth, and it grieved His heart; and He said: I will wipe The Adam that I have created off the face of the Earth." The instrument of destruction was the coming Deluge.

This, then, was the "Wickedness" that troubled Enlil: The intermarriage between the sons of the gods and female Earthlings—an intermarriage not between different races of the same species, but between *two different planetary species*—a practice that Enlil, a by-the-book disciplinarian, considered an absolute taboo. He was angered by the fact that it was none other than Enki who was first to break the taboo by having sex with female Earthlings; and he was especially infuriated by the fact that Enki's son Marduk went ahead and actually took as a wife one such Earthling—setting (in Enlil's opinion) a perverted example to the rank and file Anunnaki.

There was more to it: The forbidden liaisons produced children. We continue to read in Genesis 6:

> The *Nefilim* were upon the Earth
> in those days and also after that,
> When the sons of the *Elohim*
> came unto the daughters of The Adam,
> and they bore children to them.

No wonder that the Great Disciplinarian said: "I will wipe The Adam that I had created off the face of the Earth."

* * *

Setting aside the morals or rules that should govern interplanetary visitations, the basic problem raised by these Mesopotamian/biblical tales of our origins is this: How could the intermarried Anunnaki males and Earthling females have children—a result from mating that requires astounding genomic compatability, especially in the X (female) and Y (male) chromosomes? Indeed, taking the puzzle to its beginnings—how could the wild hominid of the Abzu have the same DNA that the Anunnaki had, similar enough so that just a little genetic mixing produced a Being that, according to the Sumerians and the Bible, was akin to the 'gods' both inwardly and outwardly except for their longevity?

The puzzle deepens by the fact that not only human, not only mammalian, not only all animal—but all life on Earth, from birds to fishes, flora to algae, and down to bacteria and viruses—all have the very same

DNA, the four nucleic acid 'letters' from which all genes and genomes are made up. That means that ***the DNA of the Anunnaki matched the DNA of all life on Earth***. And if—as should be assumed—the DNA of the Anunnaki was the same as the DNA of all life on Nibiru, then ***we must conclude that the DNA on Planet Earth and the DNA on Planet Nibiru were the same.***

How could that be, if according to the dominant modern scientific theory the Earth's seas served as a mixing bowl in which basic chemical molecules, bumping into each other and heated by geysers, somehow combined into living cells. The nucleic acids that combined to form DNA—modern scientists explain—had come about as a result of *random* bumping of chemical molecules in some *random* primordial watery 'soup' until the first living *random* cell happened. But if so, then the random result here had to be different from the random result elsewhere, for no two planets or even moons in our own one solar system are identical, and the odds that the random outcome would nevertheless be identical are virtually nil. So how did Life on Earth begin if it is so similar to Nibiru's?

The answer was given in the very tale of the Celestial Battle, when (in the second round) Nibiru/Marduk "trod upon"—came into actual contact with—Tiamat, severing her 'veins' and thrusting away her 'skull'—the future Earth. It was then that the "SEED OF LIFE"—the DNA of life on Nibiru—was transferred to Planet Earth.

Science's 'Primordial Soup' theory—whether or not valid in respect to any planetary environment elsewhere—runs into acknowledged additional problems when it comes to Earth. Abandoning the notion that the Solar System has not changed a bit since it began to take shape some 4.5 billion years ago, ***modern science now acknowledges that something extraordinary happened about 3.9 billion years ago***. In the words of *The New York Times* ('Science Times' of June 16, 2009),

> Some 3.9 billion years ago, a shift in the orbit of the Sun's outer planets sent a surge of large comets and asteroids careening into

> the inner solar system. Their violent impacts gouged out the large craters still visible on the Moon's face, heated Earth's surface into molten rock and boiled off its oceans into an incandescent mist.
>
> Yet rocks that formed on Earth 3.8 billion years ago, almost as soon as the bombardment had stopped, contain possible evidence of biological processes.

The impossibility of life starting here in such circumstances, the *New York Times* stated, has frustrated researches so much that

> Some scientists, as eminent as Francis Crick, chief theorist of molecular biology, have quietly suggested that life may have formed elsewhere before seeding the planet.

The theory that life on Earth was "***seeded from elsewhere,***" known as the Panspermia Theory, was fully discussed in my 1990 book *Genesis Revisited,* where it was of course pointed out that the 'inexplicable catastrophic event' 3.9 billion years ago was the tale of Nibiru and the Celestial Battle. The 'Panspermia' solution is neither "quietly held" (though not adopted by the scientific establishment, its proponents include many prominent scientists) nor is it new—it was put forth in cuneiform clay tablets millennia ago . . . Life on Earth and life on Nibiru—DNA on Earth and DNA on Nibiru—is the same because the ***Seed of Life*** was imparted by Nibiru to Earth during the Celestial Battle. The obtainment of such a ready-made Seed of Life explains how life could begin on Earth in the relatively immediate aftermath of the cataclysm.

Since Nibiru, at the time of the collision, already possessed formed DNA, evolution began there much earlier than on Earth. One cannot say how much earlier; but in terms of 4.5 billion years, just 1 percent earlier would mean a head start of 45,000,000 Earth-years—more than enough evolutionary time for Nibiru's astronauts to meet a *Homo erectus* on Earth.

* * *

The ancient notion that Life on Earth began when it was 'seeded' from/by Nibiru was further expressed in the concept of an actual Seed of Life—***Numun*** in Sumerian, *Zeru* in Akkadian, *Zera* in Hebrew. That basic scientific idea not only explained *how* Life on Earth originated—it also pointed to *where* on Earth life began.

It is noteworthy that in *Genesis* (1:20–25) the Bible describes the *evolution* of "Living Things" (on the *Fifth Day* of Creation) as proceeding from the waters to dry land, progressing from "all that creeps in the waters" through amphibians to the "great lizards" (dinosaurs), followed by birds, and then to "all other living creatures after their kind"—a veritable ancient Theory of Evolution whose sequence is in impressive accord with modern theories of Evolution (including the most recent findings that birds evolved from dinosaurs).

But when it concerns where Life on Earth *started,* the Bible precedes marine life with an earlier phase: On *Day Three,* according to the Bible, Life began with the appearance of *seed-bearing grasses on dry land.* It was after the formation of raised continents and water-filled seas that God said (verses 1:11–13):

> Let the Earth bring forth grass
> —the herb yielding seeds—
> and the fruit tree that yields fruit after its kind,
> whose seed is in itself;
> and it was so upon the Earth.
>
> And the Earth brought forth grass,
> and herb yielding seed,
> and the fruit tree yielding fruit
> whose seed was in itself, after its kind.
>
> And God saw that it was good;
> And the evening and the morning
> were the Third Day.

So, while in other verses *the Bible describes Evolution as we know* it, from primitive marine to fishes to amphibians, reptiles, birds, and mammals—the Bible also asserts that before "all that creepeth" began to stir in the waters, ***herbage bearing and stemming from seeds*** was the first phase of Life on Earth.

Such a distinction between the *evolution* of Life and the *start* of Life on Earth has long been held as contradicting modern science—until the publication, in July 2009 (*Nature* No. 460), of a revolutionary study according to which *"a thick, green carpet of photosynthetic life exploded across the Earth" hundreds of millions of years* ***before*** *life with "oxygen hungry cells" appeared in the waters.* Earth, the scientific journal announced, ***was "greened over" with a "thick carpet of plant life"*** whose sediments, when washed into the oceans, may have nourished watery life.

These new revolutionary findings restate what was stated in the Bible millennia earlier.

This sequence, the Bible makes clear, was made possible by the "seed" aspect of the grasses. The words 'seed,' 'seeds,' 'bearing seeds' are repeated six times in the two quoted verses, making sure that the reader does not miss the point: **Life on Earth began with/from a seed of ready-made DNA**.

Though a parallel specific Mesopotamian text has not been found thus far, other clues indicate that such a sequence of life's beginnings from herbal seeds had been noted by the Sumerians. We find the evidence in the words and terminology of the Fifty Divine Names that were granted to Marduk when he assumed supremacy. Retained in their original Sumerian form even in the Babylonian text, each name was followed by text-lines elaborating its meaning. Of immediate relevance to our subject are the following seven epithet-names; we list them as they appear in the tablet, together with their textual elucidations:

> **Maru'ukka**, Verily the god Creator of All.
> **Namtillaku**, The god who sustains life.
> **Asaru**, Bestower of cultivation,
> creator of herbs and grains
> who causes vegetation to sprout.

> **Epadun**, Lord who sprinkles the field . . .
> who establishes seed rows.
> **Sirsir**, Who heaped up a mountain over Tiamat . . .
> whose 'hair' is a grain field.
> **Gil**, Who heaps grain in massive mounds,
> who brings forth barley and millet,
> who furnishes the Seed of Earth.
> **Gishnumunab**, Creator of the Primeval Seed,
> the seed of all people.

The above sequence of attributes conforms to the Anunnaki's theory of both the origin of Life on Earth and its evolutionary stages. According to it, the celestial Marduk (alias Nibiru) is (a) the "Creator of the Primeval Seed," (b) who "furnished the Seed of Earth," beginning with herbs and vegetation that sprouts, and (c) culminating with providing "the Seed of All People." It is a notion of all life stemming from the same 'seed'—the same DNA—in a chain leading from Nibiru's "Primeval Seed" to the "Seed of All People."

In this concept—a scientific conclusion of the Anunnaki—lies the centrality of their preoccupation with "seed" as the essence of life. When Enlil wished to have Mankind perish in the Deluge, it was the "*seed of Mankind*" that Enlil wished to destroy. When Enki revealed the secret of the Flood to Ziusudra, he told him that "A Deluge will be sent to destroy the seed of Mankind." And it was not actual pairs of all animals that Noah/Utnapushtim took on board the Ark; in addition to some sheep and birds it was the "*seed of living things*" (provided by Enki) that was taken aboard. As stated in the Epic of Gilgamesh, those were the instructions to Utnapishtim:

> Man of Shuruppak, son of Ubar-Tutu,
> Tear down the house, build a ship!
> Give up possessions, seek thou life!
> Forswear belongings, keep soul alive!
> Aboard ship take thou ***the seed*** of all living things.

In the list of Fifty Names, Marduk's epithets with the term "seed" in them ranged from "He who establishes seed rows" to he "who furnishes the Seed of Earth," "Creator of the Primeval Seed," and of "the seed of all people." We can still hear the reverberating outcry of Ea/Enki—"I am the leader of the Anunnaki, ***engendered by fecund seed,*** the firstborn son of divine An!" And we must recall Enlil's superseding claim to the Right of Succession: The fact that because his mother, Antu, was a half-sister of Anu, Enlil's "seed" was doubly fecunded.

So, of whose 'seed' is Man?

The issue of our genetic origins is no longer a subsubject of biblical studies. It has moved from the realms of faith and philosophy to the arena of sophisticated science, for the latest research is zeroing in on the seemingly immortal cancer cells and the obviously fundamental stem cells (the embryonic cells from which all other body cells evolve).

In the biblical narrative, humanity stems in direct lineage from Adam (and Eve) and their son Seth through the sole surviving family of Noah and his three married sons; but even the Bible acknowledges the existence of another human lineage, the Line of Cain, that flourished in some faraway Land of Nod. Judging from the Sumerian and Akkadian sources, the actual story is considerably more complex—and it touches upon the issue of Life, Longevity, and Mortality. ***Above all, it involves the demigods—offspring of the taking by the gods of the Daughters of Man as wives.***

ADAM'S ALIEN GENES

In a historic breakthrough, two scientific teams announced in February 2001 the sequencing of the complete human genome. The principal finding was that our genome contains not the anticipated 100,000–140,000 genes (the stretches of DNA that direct the production of amino-acids and proteins) but less than 30,000—only about double the 13,601 genes of a fruit fly and barely 50 percent more than the roundworm's 19,098. Moreover, there was hardly any uniqueness to the human genes. They were found comparative to almost 99 percent of the chimpanzees, and to 70 percent of the mouse. Human genes, with the same functions, were found to be identical to genes of other vertebrates, as well as invertebrates, plants, fungi, even yeast.

The findings not only confirmed that there was *one source* of DNA for all life on Earth, but also enabled the scientists to trace the evolutionary process—how more complex organisms evolved, genetically, from simpler ones, adopting at each stage the genes of a lower life form to create a more complex higher life-form—culminating with *Homo sapiens*.

It was here, in tracing the *vertical* evolutionary record contained in the human and other analyzed genomes, that the scientists ran into an enigma. The "head-scratching discovery," as the journal *Science* (issue No. 291) termed it, was that ***the human genome contains 223 genes that do not have any predecessors on the genomic evolutionary tree***. In fact, these 223 genes were found to be completely missing in the whole range of the vertebrate phase of evolution. An analysis of the functions of these genes, published in the journal *Nature* (issue No. 409), showed that they involve important physiological and cerebral functions peculiar to humans. Since the difference between Man and Chimpanzee is just about 300 genes, those 223 genes make a huge difference.

How did Man acquire such a bunch of enigmatic genes? The scientists could only explain the presence of these alien genes by a "rather recent" (in evolutionary time scales) "probable *horizontal transfer from bacteria*," suggesting that these are not genes acquired through evolution, but genes acquired through recent ***infection from bacteria***.

If one accepts the "horizontal bacterial insertion" explanation, I wrote in my website, then it was *a group of bacteria* that said, "Let us fashion The Adam in our image" . . .

I still prefer the Sumerian and biblical Anunnaki/Elohim version.

IX
Gods and Other Ancestors

We shall never know the name—if he ever had one—of the hominid whose Ti.it was used by Ninmah in the genetic mixture for creating the gods' Workman; with the repeated trial and error, more than one hominid were involved. But we do know—due to additional cuneiform tablet discoveries—whose godly 'essence' or bloodline was used in the process.

Does it matter? Perhaps not much, in view of the varied other genealogical and genetic ancestors that Man on Earth had in the course of time. But if some genes never die, then the issue is of interest—at least from a *What If* aspect—since Mankind's record, from the very biblical beginnings, is not a happy odyssey. It is a tale more heartbreaking than ever conceived by a Shakespeare or a Homer: A wondrous creation, 'The Adam' is really fashioned to be a serf; placed in a bountiful Eden, his stay is cut short by disobeying God. Enabled to procreate, Adam is doomed to eke out a living from parched soil and Eve is condemned to give birth in agony. They bear two sons, and there are four humans on Earth; then Cain (a tiller of the land), jealous of Abel (a shepherd), slays his brother, reducing Mankind to three . . .

Serfdom, disobeyance, fratricide—are they part of our genetic makeup because we are mostly heirs to the DNA of Earth's animal kingdom—or because the bloodline selected by the Anunnaki—the 'Alien Genes'—was that of a young rebel who incited his crewmates to *kill Enlil?*

While in some texts—including references to the Creation of Man in the Epic of Creation—the god whose blood was used is executed for being the rebels' leader, other Atra-H̲asis versions explain the choice as due to that god having the right *Te'ema,* translated 'Life's Essence' or 'Personality' (genetically speaking). Where not entirely missing, the cuneiform signs giving his name used to be read (in Akkadian) *Wéila;* new tablet discoveries in the 1990s in Sippar by Iraqi archaeologists clearly name him ***Alla*** in Akkadian and **Nagar** in Sumerian—an epithet-name meaning 'Metal-craftsman', *specifically in copper.* This could suggest a deliberate choice (rather than mere punishment) in view of the fact that the *Nachash* Serpent/Knower of Secrets in the Bible's Garden of Eden tale also stems from the same verb-root as *Nechoshet,* which means *copper* in Hebrew. The fact that Nagar (and his spouse **Allatum**) are listed among the Enki gods in the various God Lists, reinforces his role as leader of the insurrection against Enlil.

Biblical scholars agree that the context for the Cain-Abel incident is the unending and universal conflict between farmers and herdsmen over land and water. Such conflicts are described in Sumerian texts as part of Mankind's early history—a theme expounded upon in a text scholars call *The Myth of Cattle and Grain,* where Enlil is the deity of ***Anshan*** (grains and farming) and Enki of ***Lah̲ar*** ('woolly cattle' and sheepherding)—roles that were continued by Enlil's son **Ninurta** who (as depicted on cylinder seal VA-243, Fig. 51) gave Mankind the plow, and Enki's son **Dumuzi**, who was a Shepherd. As in other instances, the Bible combined the two deities (Enlil and Enki) into a sole 'Yahweh' who accepts the shepherd's (Abel's) offering from his flocks but ignores the farmer's (Cain's) offering of "fruit of the soil."

Following the Cain-Abel tale, the Bible devotes the remainder of Genesis chapter 4 to Cain and his descendants. Fearful of getting killed for his sin, Cain is granted by God a visible protective 'mark' (the favorite "Mark of Cain" of Sunday preachers) that will last for "seventyfold" generations. (If transmittable through the generations, it had to be a genetic marker.) As in the tale of the Deluge, the same Yahweh who has had it with Mankind and seeks its elimination but then proceeds to save it through Noah, so does 'Yahweh' who ignored, condemned, and punished Cain now grants him safety and protection. Once again, we see, the Bible combined actions of Enki with actions of Enlil into one divine entity called "**Yahweh**." As explained to a questioning Moses (Exodus 3:14), the name meant "I will be whoever I will be"—a universal God once acting through/as Enlil, another time through/as Enki, or in time through other entities ('gods') as His emissaries.

Protected by a sympathetic deity, the wandering Cain reached "the Land of *Nod,* eastward of Eden." There Cain "knew his wife" and had a son, ***Enoch*** (= 'Founding' or 'Foundation'); and he built a city, and named it 'Enoch' in honor of his son. Then "unto Enoch ***Yared*** was born, and Yared begot ***Meẖuyahel;*** and Meẖuyahel begot ***Metusha'el,*** and Metusha'el begot ***Lamech.***"

On reaching the seventh generation (Adam-Cain-Enoch-Yared-Meẖuyahel-Metusha'el-Lamech), the Bible gets generous—even praiseful—with its information on the Cain line and its achievements:

> And Lamech took unto himself two wives,
> the name of one Addah and of the other Zillah.
> And Addah bore ***Jabal;*** he was the father of such
> that dwell in tents and have flocks;
> and his brother's name was ***Jubal***—he was the
> father of all who play the harp and pipe.
> And Zillah too gave birth, to ***Tubal-Kain***—
> an artificer of every article of copper and iron.
> And the sister of Tubal-Kain was ***Na'amah.***

These accomplishments of seven generations in the Cain lineage were celebrated by Lamech with a song; quoted by the Bible, it combined Cain's "seventyfold" with an invoking of an enigmatic "seventy-seven" by Lamech, to form a symbolic Triple Seven (7-7-7).

In spite of its brevity, the Cain Line tale in the Bible depicts a high civilization that started with a toiler of the land, passed through a Bedouin-like stage of nomadic tent dwellers who tend flocks, and mastered a transition from peasantry to city dwelling, boasting musicians and including metallurgists. If not in the pre-Diluvial Edin or in the *future* Sumer, where did such a civilization arise?

The Bible avoids telling us where Cain settled, stating only that he went to the "east of Eden," toward "the land of *Nod*" (= 'Wandering'). We are left guessing how far Cain went to the "east of Eden"—just to the lands of the Zagros Mountains that later on became Elam, Gutium, and Media? Did he and his family keep wandering eastward on the Iranian plateau, to the metalworking land of Luristan and the cattle-rich Indus Valley? Did these wanderers reach the Far East? Did they, perhaps, even cross the Pacific Ocean, reaching the Americas?

It's not an absurd question, since Man did, somehow, sometime in the early past, reach the Americas—thousands of years before the Deluge. The puzzle is Who, How, and When.

The general scholarly assumption has been that the Sumerians (and their successors in Mesopotamia) had no interest in, and thus no record of, a 'lost line' of Cainites. But it is inconceivable that the biblical section about Cain's migration, generations, and their impressive achievements was not based on some Mesopotamian written record. In fact, such a very tablet, now archived in the British Museum (catalogued BM 74329)—transcribed (Fig. 68), translated and reported by A. R. Millard and W. G. Lambert in the journal *Kadmos* (vol. VI)—speaks of a group of exiled people who were "plowmen" (as Cain was, "a tiller of the land"). They wandered and reached a land called *Dunnu* (the Bible's 'Land of *Nod*'?); there their leader, named *Ka'in* (!), built a city whose landmark was a twin tower:

> He built in Dunnu
> a city with twin towers.
> Ka'in dedicated to himself
> the lordship over the city.

The clue about a city noted for its twin towers is especially intriguing. Early human arrival in the Americas via the Pacific Ocean is not only the latest scientific conclusion, but is in accord with local native lore both in South and North America. In Mesoamerica, the legendary arrival by

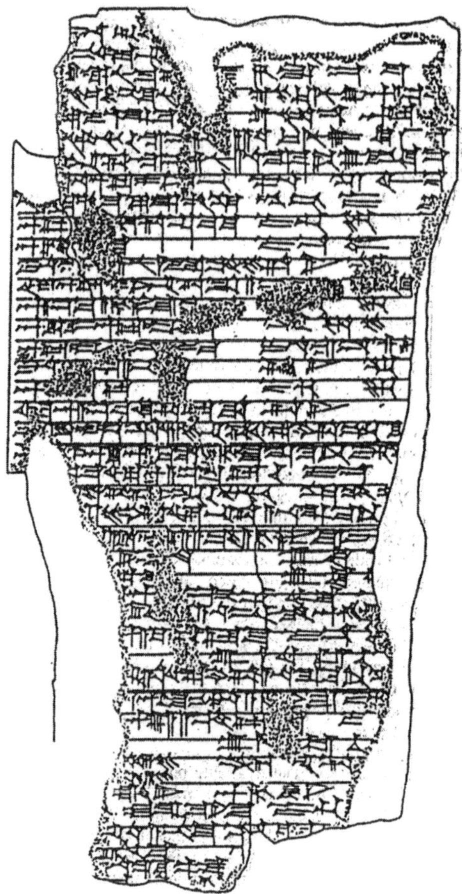

Figure 68

boats was from an ancestral land of *Seven* Caves or *Seven* Shrines (Fig. 69, from a pre-Aztec Nahuatl codex). Pointing out the parallels with the 7-7-7 in the Cain/Lamech line, I wondered in *The Lost Realms* and in *When Time Began* whether the name of the Aztec capital, *Tenoch-titlan* (= 'City of *Tenoch*'), now Mexico City, might have really meant 'City of *Enoch*', **a city known when the Spaniards arrived by its twin-towered Aztec temple** (Fig. 70). I also speculated whether the 'Mark of Cain', which had to be noticed and recognized by others on sight, could have been the Mesoamerican men's absence of facial hair.

Figure 69

The text's similarities to the biblical story of Cain's wanderings and the city he built are obvious—but the presumption is that all that took place within the geographic embrace of the Near East. A transpacific jump to the Americas nevertheless refuses to vanish, because the detail of *four brothers marrying their sisters and founding a new city* is the core of the main Legends of Beginnings of the native peoples of South America. There (as detailed in *The Lost Realms*) the legend was of the *four Ayar Brothers* who married their sisters, went wandering, and founded the great city of Cuzco with its temple; they found the correct site for this "Navel of the Earth" with the aid of a golden wand given them by the god Viracocha (= 'Creator of All').

Figure 70

As one remains confounded by these similarities, one thing can be asserted with certainty: If the legends (and the people) traveled, it was from the Near East to the Andes, not vice versa. If that is how it happened, then we have here a segment of Mankind that might have survived the Deluge without Noah's ark, offering a human genetic lineage without the intermarriage intrusion.

* * *

Without a pause, the Bible follows the Lamech/777 verses with the information that back home, "Adam knew his wife again, and she bore a son, and she called his name ***Sheth***"—'Seth' in English—a wordplay name meaning in Hebrew 'Granted', "for God hath granted another seed instead of Abel whom Cain had slain." Seth, let it be noticed, is not just another son—he is "another seed." "And to Seth, to him also a son was born, and he called his name ***Enosh;*** it was then that calling by the name of Yahweh began" (Genesis 4:26).

The Bible's words thus make clear that with the birth of Enosh to Seth, a new genealogical/genetic line has been launched; it leads straight to Noah and thus to the post-Diluvial surviving 'Seed of Mankind.'

The name ***Enosh*** is not difficult to explain: It means, in Hebrew, "Human" in the sense of 'One Who Is Frail/Mortal'. It stems from the same root as the term *Enoshut,* and undoubtedly coming from the Akkadian *Nishiti;* both mean "Humanity, Humankind"; and it is clear that it is this human lineage (as distinct from the one through the exiled Cain) that is involved in the ensuing events, ***including the intermarriage with the sons of the Elohim.***

The Bible's emphasis on this genealogical line is expressed by the 'editorial' placement and extent of the data. The line of Cain is described in eight verses, inserted in chapter 4 of Genesis between the story of Cain and Abel and the birth (to Adam and Eve) of Seth and Enosh. To the line through Seth and Enosh, the Bible devotes the two concluding verses in chapter 4 plus all of chapter 5 and its 32 verses. The list provides an uninterrupted genealogical chain of ten pre-Diluvial Patriarchs from Adam to Noah, leaving no doubt that it is this lineage that led to

Noah and thus to the salvaging of the Seed of Mankind and its restoration in the aftermath of the Deluge.

Though it is the favored genealogical line, the Bible is stingy with information about it. With one main exception, the data that the Bible provides consists of a name, at what age each Patriarch gave birth to his firstborn son, and how long he lived thereafter. But who were they, by what were they distinguished, what were their occupations? The only aspect of their lives that is evident right off is that they were blessed with impressive longevity:

> ***Adam*** lived 130 years and begot a son in his likeness
> and after his image, and called his name Sheth.
> And the days of Adam after he had begotten Sheth were
> 800 years; and he begot [other] sons and daughters.
> [So] all the days that Adam lived were 930 years,
> and he died.
>
> And ***Sheth*** lived 105 years and begot Enosh.
> And Sheth lived after he had begotten Enosh 807 years,
> and he begot sons and daughters.
> And all the days that Sheth lived were 912 years,
> and he died.

The list continues in the same manner for the next four Patriarchs—***Enosh*** begets Kenan at 90, lives another 815 years, begets other sons and daughters, dies at age 905. ***Kenan*** begets Mahalalel at 70, dies at 910; ***Mahalalel*** begets Yared ('Jared' in English) at 65, dies at age 895; ***Yared*** begets Enoch at age 162, dies at age 962.

There was an extraordinary occurrence when it came to the seventh Patriarch, ***Enoch,*** who "lived 65 years and begot Methuselah," but did not die because, at age 365, "*Elohim* had taken him." We shall return to this significant revelation shortly; right now we shall continue the record of the subsequent Patriarchs to complete their list and their age counts:

> ***Methuselah*** gave birth to Lamech at age 187 and died at
> age 996.

Lamech gave birth to Noah at age 182, died at age 777.
Noah gave birth to Shem, Ham, and Japheth at age 500;
he was 600 years old when the Deluge swept over the Earth.

While on the face of it these numbers indicate remarkable longevity (which is expected of those closer to the genetic infusion), the list suggests that the Patriarchs lived to see not just children and grandchildren, but also great-grandchildren and great-great-grandchildren and beyond—and died just ahead of the Deluge. Thus, in spite of their extraordinary longevities, it took a mere 1,656 years from Adam to Noah:

	Adamic Years	
Adam born	**0001**	
Seth born	130	
Enosh born	235	
Kenan born	325	
Mahalalel born	395	
Yared born	460	
Enoch born	622	
Metushelah born	687	
Lamech born	874	
Adam dies	**930**	(age 930)
Enoch transfigured	987	(age 365)
Seth dies	1042	(age 912)
Noah born	**1056**	
Enosh dies	1140	(age 905)
Kenan dies	1235	(age 910)
Mahalalel dies	1290	(age 895)
Yared dies	1422	(age 962)
Noah's 3 sons born	1556	
Lamech dies	1651	(age 777)
Metushelah dies	1656	(age 969)
Flood (Noah 600)	**1656**	

Odd or not, this lineage list of ten pre-Diluvial Patriarchs, leading to the hero of the Deluge and to the story of the Deluge, unavoidably invited scholarly efforts to compare it with the ten ante-Diluvial kings of Berossus and his sources—not an easy task, since the Bible's mere 1,656 years from Adam's birth to the Deluge is quite different from the 432,000 years of Berossus (or the totals according to WB-62, WB-444, etc.):

Bible		WB-62		Berossus	
Adam	130	Alulim	67,200	Aloros	36,000
Seth	105	Alalgar	72,000	Alaparos	10,800
Enosh	90	[En]kidunu	72,000	Amelon	46,800
Kenan	70	[. . .]alimma	21,600	Ammenon	43,200
Mahalalel	65	Dumuzi	28,800	Megalaros	64,800
Yared	162	Enmeluanna	21,600	Daonos	36,000
Enoch	65	Ensipzianna	36,000	Euedorachos	64,800
Metushelah	187	Enmeduranna	72,000	Amempsinos	36,000
Lamech	182	Sukurlam (?)	28,800	Ardates (or Obartes)	28.800
Noah	600	Ziusudra	36,000	Xisuthros	64,800
Ten Patriarchs	1656	Ten rulers	456,000	Ten kings	120 *Shars* = 432,000

There have been numerous scholarly attempts of numerical gymnastics aimed at finding some common denominator between the 1,656 years and the Mesopotamian numbers; none are convincing or reasonably acceptable. Our own attempt (in *Divine Encounters*), focusing on the obvious Noah/Ziusudra identity and thus the 600:36,000 relationship, pointed out that since the numeral "1" in cuneiform could also mean "60" depending on its position, it could well be that the biblical redactor reduced the ages by a factor of 60. That would mean a span of 99,360 (1,656 x 60) years from Adam to Deluge—still not enough to close the gap.

That the numbers don't add up is no wonder, for the usual computing method is wrong to begin with. The Mesopotamian count begins with the arrival of the Anunnaki (120 *Sars* before the Deluge); the Adamic count should begin not from the same moment, but from the time of fashioning The Adam—40 *Sars* later—and even later still, from when the individual called 'Adam' was born. Furthermore, the Mesopotamian list gives the lengths of reign, which at best should be compared to when Patriarchal succession took place, not when a son was born.

Using life-span figures rather than birth-of-son dates, and multiplying those ages by 60, results in a better 'Berossus-like' range: Adam's 930 would become 55,800 years, Seth's 912 will be 54,720, the 905 of Enosh 54,300, and so on. Added together, the ten life spans (with Enoch's count stopped at 365 and that of Noah's at 600) come to a grand total of 8225, which multiplied by 60 results in 493,500 years. Assuming that succession sometimes took place before the predecessor's death, we come within range of the Mesopotamian totals.

A better track worth following might be comparing personalities, using their names and/or occupations as clues. Could we find, for example, the point in the Mesopotamian ten-kings lists where the biblical Adam makes an appearance? It seems that we can, if we look carefully.

Of the first two rulers, we definitely know that they reigned in Eridu, the first Anunnaki settlement established by Ea/Enki. Both bore typical early 'Anunnaki' names; in all probability, **Alulim** was **Alalu**, the deposed Nibiruan king, appointed Chief Administrator ('king') in Eridu by his son-in-law Ea/Enki. **Alalgar**, whose name conveyed the notion of 'settling down', is not otherwise known, and could have been one of Enki's aides.

The interesting point about their reigns, as recorded in WB-62, is that together they totaled 139,200 years—just under the 40 *Shars* (= 144,000 years) of Anunnaki toil before 'Workman' was fashioned. It seems as the right moment for The Adam, born to toil, to appear. And indeed, it is here that the Mesopotamian list names third ruler ***Amelon***—"The Workman" in Akkadian—a rendering that matches the Sumerian *Lulu-**Amelu***. Looking at his name in the WB-62 list, the answer stares right into our eyes: ***Enki.dunnu*** simply and clearly means in Sumerian "***Enki made/fashioned him.***"

In the Akkadian '*Amelon*' and the Sumerian 'Enki.dunnu', I suggest, we are staring at the biblical 'Adam'.

WB-62 then lists two names: The incomplete ***[. . .]-Alimma*** and "***Dumuzi,*** a shepherd." The names and their sequence give us pause; incredibly, **Alim** means, in Sumerian, 'Grazing land' or its animal, the ram; **Dumu.zi** literally means "Son who is Life." Could these Sumerian names stand for Adam's sons Abel, the Herder, and then Seth, the son through whom new Line of Life was granted?

Various studies comparing the biblical list of Patriarchs with the Berossus list have already suggested that *Ammenon* in Berossus stems from the Akkadian (and Hebrew) term for craftsman/artificer, *Amman*—a description befitting the biblical *Kenan* (= 'Artificer of Implements'). Without dwelling on the rest of the names, the instances thus far given strongly suggest *one common source* for the various Sumerian King Lists, Berossus, and the Bible.

Our analysis and findings go beyond the conclusion that somewhere, somehow, there had to be a common source from which the data was obtained. For if the Sumerian pre-Diluvial rulers and the biblical pre-Diluvial Patriarchs were the same, it raises the question: Who, indeed, were these Patriarchs? If Adam and Seth and Enosh, etc., lived and 'reigned' for periods counted in *Shars,* could they have been mortal men (as the Bible implies)? If they were the *Shar*-span rulers of the Sumerian King List, why the repeated biblical statement that each one of them died? Or were they perhaps a combination of the two: partly mortal men, partly gods—in other words, ***Demigods***—with all the genetic consequences thereof?

Could the biblical Patriarchs themselves, including Noah, have been the very "Men of Renown" of Genesis chapter 6 who were fathered by the *Nefilim* who had mated with the 'Daughters of Man'?

For an answer—an amazing answer—we have to take another look at all the available sources.

THE POWER OF SEVEN

Our daily life is regulated by the *seven*-day week—an odd number that fits neither our decimal (= 'Base ten', as the number of our digits in two hands) system, nor the Sumerian sexagecimal (= 'Base Sixty') system that we continue to use in geometry, astronomy, and timekeeping. This unusual choice is explained by the biblical tale of Creation that covered the span of *seven* days (the final day of rest and review included). This biblical seven is explained in turn by the *seven* tablets of *Enuma elish,* the Mesopotamian Epic of Creation. But why is that text inscribed on seven tablets?

The number seven (including seventh and seventy) appears in almost every major biblical event, commandment, and prophecy, for a total of some six hundred times. It is also a key number in the New Testament, including the prophetic *Book of Revelation,* as well as in the Pseudoepigraphic books (such as the *seven* classes of angels in the *Book of Enoch*).

That has been the same in Egyptian lore, starting with the affairs of the gods: The first divine dynasty consisted of *seven* gods (from Ptah to Horus); and in all there were 49 (= 7 x 7) divine and demigod rulers until Pharaonic reign began. Mesoamerican beginnings are attributed to seven tribes; and so on.

The consideration of *seven* as a Power Number in fact began with the Anunnaki who had come to Earth from Nibiru. Nippur, Mission Control Center, was the seventh city on Earth. There were seven Sages, and the 'Seven Who Judge'. Ziggurats had seven stages, and stars were located with the "stylus of seven numbers." A god had the 'Sevenfold Weapon', and there were seven 'Weapons of Terror'. The release of the Bull of Heaven triggered seven years of famine; when a temple was inaugurated, seven blessings were pronounced. And so on and on.

The origin of all that, we suggest, is the position of Earth as the seventh planet from the viewpoint of the Anunnaki (see the sky map of Enlil's route from Nibiru to Earth, Fig. 65). It states that "Enlil went by *seven planets*" to reach Earth—starting the count with Pluto, then Neptune and Uranus as 2nd and 3rd, Saturn and Jupiter as 4th and 5th, Mars as 6th, and ***Earth as the Seventh Planet***. Accordingly, *seven dots* was Earth's celestial symbol, as seen on an Assyrian monument (alongside the symbols for the Moon, Nibiru, and the Sun, and the deities associated with them).

X

Of Patriarchs and Demigods

A 'demigod', by definiton, is someone who is a product of the mating of a god (or goddess) with an Earthling, sharing the two genomes. As startling, or dismissed as myth, as the possibility may sound, the Bible unambiguously asserts that such mating had taken place, and that heroic "Men of Renown" were born as a result both before and after the Deluge. On the face of it, that is all the Bible has to say on such a history-changing matter (it was the cause for the plan to terminate Mankind by the Deluge!)—unlike the Mesopotamian texts that are filled with tales of demigods, with Gilgamesh notorious among them. And that, as we shall see, opens the door to potential discoveries in our own present time.

Some probing of available material, enhanced with deductive reasoning, will show that the meager biblical data about the pre-Deluvial Patriarchs dovetails with the more extensive Mesopotamian information. The brief biblical statement in Genesis 6 about the "sons of the *Elohim*" who had taken Daughters of Man as wives is also substantially augmented in other ancient Hebrew writings—'Lost Books' that have not made it into the canonical Hebrew Bible—collectively known as *Apocrypha* (= 'Secret, Hidden Writings') or *Pseudo-Epigrapha* of the Old Testament; and it behooves us to explore that too.

That such writings existed is confirmed by the Bible itself; it refers to several 'lost books' whose existence (and contents) were common knowledge at the time, but that have since been lost. Verse 14 in

Numbers 21 refers to the *Book of the Wars of Yahweh;* Joshua 10:13 recalls the miraculous events described in the *Book of Yasher.* Those, and other mentioned books, have been completely lost. On the other hand, some lost books—such as *The Book of Adam and Eve, The Book of Enoch, The Book of Noah,* and the *Book of Jubilees* have come down to us through the ages preserved by translations in languages other than Hebrew, sometimes partly or entirely rewritten by the later renderers. These manuscripts are important not only for reiterating biblical data, but also because they purport to provide added details to biblical tales; and some of them record the incident of the intermarriage and fill in the details.

The Bible, in Genesis 6, presents a God who is of two contradictory minds. He is angered by the intermarriage of the "sons of the *Elohim*" with the Daughters of Man, yet later on considers the offspring to be heroic "Men of Renown." He decides to wipe mankind off the face of the Earth, then goes out of the way to save the Seed of Mankind via Noah and the ark. We now understand that the apparent contradictions stem from the combining by the Bible of diverse and opposed deities, such as Enki and Enlil, into one divine entity (*Yahweh*). The authors of the *Book of Jubilees* and the *Book of Enoch* dealt with the duality problem by explaining that the descending of the Angels to Earth was meant to be benevolent, but then a group of them were led astray by errant leaders to take Earthlings as wives.

It happened, the *Book of Jubilees* reported, during the time of ***Yared*** (= 'He of Descending') who was so named by his father, Mahalalel, because it was then that the "Angels of the Lord descended to the Earth." Their mission was to "instruct the children of men with judgment and uprightness"; but instead they ended up "defiling themselves" with the Daughters of Man.

According to those extra-biblical texts, some two hundred 'Watchers' (= the **Igigi** of Sumerian lore) organized themselves in twenty groups of ten; each group had a named leader; most of the names—Kokhabiel, Barakel, Yomiel, etc.—are theophoric names honoring *El* (= Lofty). One, called Shemiazaz, who was in overall command, made all of them swear to act together. Then "each one of them chose for himself one, and they began to go in unto them and defile themselves with them . . . ***And the women bore giants.***"

But according to the *Book of Enoch,* the instigator of the transgression, "the one who led astray the sons of God and brought them down to Earth and led them astray through the Daughters of Man," was actually the wrong-doing angel *Azazel* (= 'The Might of *El*') who was exiled for his sins. According to Mesopotamian texts, which include segments dealing with Marduk's exile, Marduk was the first one to break the taboo and marry (as distinct from just having sex) Sarpanit, an Earthling woman, and to have a son (named Nabu) by her; and one is left wondering to what extent Marduk's involvement played a role in Enlil's anger.

* * *

Enoch, it will be recalled, was the next pre-Diluvial Patriarch after Yared who "walked with the *Elohim*" and did not die, for he was taken away to be with them; as stated in Genesis 5:21–24:

> And Enoch walked with the *Elohim,*
> after he had begotten Metushelah,
> 300 [more] years,
> and begot [other] sons and daughters.
> And all the days of Enoch were 365 years.
> Enoch walked with the *Elohim* and was no more,
> for *Elohim* had taken him.

The book ascribed to him, *The Book of Enoch,* enlarges on that statement and describes the Watchers' affair as the reason why the Righteous Angels had revealed to Enoch secrets of Heaven and Earth, the Past and the Future: The purpose was to set Mankind, through the revelations to Enoch, on a righteous path—a path from which it was diverted by the Watchers' misdeeds.

Enoch, according to these writings, was taken heavenward twice; and whereas the Bible simply states that he first "walked with the

Elohim" and then was "taken" by them, the *Book of Enoch* describes a plethora of angels and archangels who carried all that out.

His sojourn with "the Holy Ones" began with a dream-vision in which his bedroom, he wrote later, filled with "clouds which invited me and a mist which summoned me," and a kind of whirlwind "lifted me upwards and bore me unto heaven." Miraculously passing through a fiery crystal wall, he entered a crystal house whose ceiling emulated the starry skies; then, reaching a crystalline palace, he saw the Great Glory. An angel led him closer to a throne, and he could hear the Lord tell him that he was chosen to be shown "the heavenly secrets" so that he could teach them to Mankind. He was then told the names of the seven archangels who serve the Lord and who will be his mentors on his journey of discovery. With that, his dream-vision came to an end.

Later on—exactly ninety days before Enoch's 365th birthday—as Enoch was alone in his home, "two men, exceedingly big" whose appearance "was such as I have never seen before," materialized out of nowhere. Their faces shone, their clothing was unlike any, and their arms were like golden wings. "They stood at the head of my couch, and called upon me by name," Enoch later told his sons Metuhsha'el and Regim.

The two divine emissaries told Enoch that they have come to take him on a second, prolonged celestial journey, and suggested that he inform his sons and servants that he will be gone for a while. Then the two angels took him on their wings and carried him to the First Heaven. There was a great sea there; and it was there that Enoch was taught the secrets of climate and meteorology.

Continuing the journey, he passed through the Second Heaven, where sinners were punished. In the Third Heaven was Paradise, where the righteous go. In the Fourth Heaven—the longest stop—the secrets of the Sun, Moon, the stars, the zodiacal constellations and the calendar were revealed to Enoch. At the Fifth Heaven the link bonding Earth with Heaven petered out; it was the abode of the "angels who connected themselves with women." It was there that the first part of Enoch's celestial journey was completed.

Resuming his journey, Enoch passed through the Sixth and Seventh Heavens, where he encountered diverse groups of angels, arranged by ascending order: Cherubim, Seraphim, Archangels—seven classes in all. Reaching the Eighth Heaven, he could actually see the stars that make up the constellations. At the Ninth Heaven he could see the realm of the Zodiacs.

Finally he reached the Tenth Heaven, where the angels brought him "before the Lord's face." Terrified, he fell to his knees and bowed. And the Lord spoke to him and said:

> Arise Enoch, have no fear!
> Arise and stand before my face,
> and gain Eternity.

And the Lord commanded the archangel Michael to change Enoch's earthly garments, and clothe him in divine garments, and anoint him. And the Lord told the archangel Pravu'el to "bring out the books from the sacred storehouse, and a quickwriting reed," and give them to Enoch so that he could write down all that the archangel will read to him—"all the commandments and teachings." For thirty days and thirty nights Pravu'el dictated and Enoch wrote down "the secrets of the workings of the heavens, the Earth and the seas, and all the elements . . . The thunderings of thunders, and the Sun and the Moon, the comings and goings of the stars, and the seasons, years, days, and hours." He was also taught "human things"—like the "tongues of human songs." The writings filled up 360 books. Returned to the presence of the Lord, Enoch was seated to His left, beside the archangel Gabriel; and the Lord himself told Enoch how Heaven and Earth were created.

And then the Lord told Enoch that he will be returned to Earth for thirty days so that he could bequeath to Mankind the handwritten books, to be passed from generation to generation. Returned to his home, Enoch told his sons of his odyssey, explained to them the books' contents, and admonished them to be righteous and follow the commandments.

Enoch was still talking and explaining when his thirty-day homecoming was up; by then, word having spread in town, a great crowd of people gathered around Enoch's home, striving to hear the details of the celestial journey and the heavenly teachings. So the Lord caused darkness upon the Earth; and in the darkness two angels swiftly lifted Enoch and carried him away "to the highest heaven."

Realizing that Enoch was gone, "the people could not understand how Enoch had been taken; they went back to their homes, and those who witnessed such a thing glorified God." And the sons of Enoch "erected an altar at the place where Enoch had been taken up to heaven." It happened, a scribal postscript states, exactly when Enoch reached the age of 365 years—a number alluding to his newly acquired mastery of astronomy and the calendar. (One recalls at this point the statement by Manetho regarding a dynasty of 30 demigods in Egypt who reigned a total of 3,650 years—a number that is precisely 365 x 10. A mere coincidence?)

It is noteworthy that neither the Bible in its brief information regarding Enoch, nor the 100-plus chaptered *Book of Enoch,* explain why Enoch was chosen for the extraordinary divine encounters and avoided a mortal's death; how was he special, different? The name of the one who "begot" him, Yared, is explained by the notation that it was in his time that the Descending (of the *Nefilim*) had occurred. The name *Yared* is clearly derived from the root verb meaning "To Descend" in Hebrew; but it is grammatically awkward, leaving it unclear whether Yared himself is 'One who had descended', which would grant him a god's status and make his son a demigod.

Also left untold is what was the city where Enoch lived, the locale of miraculous events and site of an altar commemorating them. If it was also the town of his father, *Yared*—the parallel of the Cainite ***Yirad***— one wonders whether the name ***echoes the city's name,*** **ERIDU.**

If so—if the site of Enoch's divine encounters was Eridu of Enki and Anunnaki fame—we have here details that link those biblical and extra-biblical pre-Diluvial Patriarchs back to the Sumerian pre-Diluvial Kings and to the "sons of the *Elohim*" (whom the Bible itself describes as *Gibborim*—heroic "Men of Renown").

* * *

The possibility that pre-Diluvial biblical Patriarchs were demigods, has loomed large already in antiquity—especially in the case of Noah.

The *Book of Enoch,* scholars have concluded, incorporated sections of another, earlier lost book—a *Book of Noah.* Its existence was surmised from various other early writings and from the different writing style of sections within the *Book of Enoch.* The surmising became a certainty when fragments of a *Book of Noah* were discovered among the Dead Sea scrolls—a virtual library that was hidden in caves at a site called Qumran on the shores of the Dead Sea in Israel some 2,000 years ago. In that scroll the word usually translated 'Watchers' clearly calls them *Nefilin* (Fig. 71)—Aramaic for ***Nefilim*** in Hebrew.

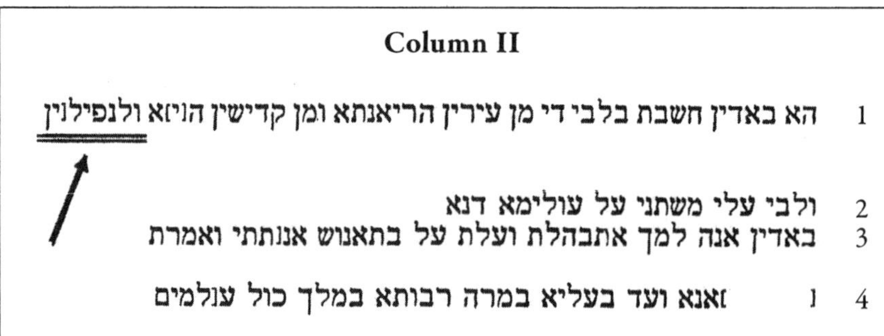
Column II

1 הא באדין חשבת בלבי די מן עירין הריאנתא ומן קדישין הו]א ולנפילין

2 ולבי עלי משתני על עולימא דנא

3 באדין אנה למך אתבהלת ועלת על בתאנוש אנותתי ואמרת

4]]ואנא ועד בעליא במרה רבותא במלך כול עולמים

Figure 71

According to the relevant sections of the book, the wife of Lamech (father of the biblical Noah) was named ***Bath-Enosh*** (= 'Daughter/Offspring of Enosh'). When Noah was born, the baby was so unusual that he aroused suspicions in the mind of Lamech: He looked decidedly different from the usual baby boys, his eyes shone, and he could speak. And right away Lamech "thought in his heart that *the conception was from one of the Watchers.*" Lamech expressed his suspicions to Metushelah, his father:

> I have begotten a strange son,
> different from and unlike Man,
> and resembling the sons of God of Heaven.
> His nature is different, and he is not like us.
> And it seems to me that he is not sprung from me,
> but from the angels.

Suspecting that the boy's real father was one of the Watchers, Lamech questioned his wife, Bath-Enosh, demanding that she swear to him "by the Most High, the Lord Supreme, the King of all the Worlds, the ruler of the Sons of Heaven," to tell him the truth. Responding, Bath-Enosh told Lamech: "Remember my delicate feelings! The occasion is indeed alarming, and my soul is writhing in its sheath!" Puzzled by the answer, Lamech again asked her to swear to tell him the truth. Bath-Enosh again reminded Lamech of her "delicate feelings"—but then, swearing by "the Holy and Great One," assured him "that this conception was by you and not by some stranger or by any of the Watchers."

Still skeptical, Lamech went to his father, Metushelah, with a request: To seek out his father, Enoch—who was taken away by the Holy Ones—and ask him to pose the fatherhood question to them. Locating his father, Enoch, "at the ends of the Earth," Metushelah related to him the Noah puzzle and conveyed to him Lamech's request. Yes, Enoch told him, in the days of my father, Yared, "some Angels of Heaven did transgress and united themselves with women, and have married some of them, and begot children by them"; but you can reassure Lamech "that he who has been born is in truth his son." The odd features and unusual talents of Noah are due to his having been chosen by God for a special destiny, as predicted "in the heavenly tablets."

Lamech accepted those reassurances; but what are we to make of the whole tale? Was Noah, after all a demigod—in which case we, his descendants, have a larger dose of Anunnaki genes than The Adam had received?

The Bible had this to say in its introduction to the Deluge story:

> These are the generational records of Noah:

> Noah was a righteous man,
> Perfect in his genealogy he was;
> With the *Elohim* did Noah walk.

If this leaves one wondering, a re-reading of the prior Nefilim verses in Genesis 6 reinforces the impression that the Bible itself left the question hanging by stating, after verse 4 about the demigods who were "the Mighty Men of old, Men of Renown," "*And* Noah found grace in the eyes of the Lord" (verse 8). It does not say "But"—the verse starts with "And" as though it was a direct continuation of the previous verses about the sons of the gods—

> They were the Mighty Men of old,
> Men of Renown, *and* (= 'as well as') Noah
> [who] found grace in the eyes of the Lord.

Read this way, Noah would have been one of the Mighty Men of Renown—a demigod whose 600 years before the Deluge telescope the 36,000 years of Ziusudra/Utnapishtim.

* * *

Sumerian texts include the story of the pre-Diluvial ***En.me.duranki*** (also called ***En.me.duranna***) whose tale is remarkably similar to that of the biblical Enoch. His theophoric name links him to the ***Dur.an.ki*** (= 'Bond Heaven-Earth'), Enlil's command center in Nippur.

A patriarch named 'Enoch', it will be recalled, appears in the Bible in both the Cain and the Seth genealogical lines. In the context of the Enki-Enlil rivalries, Enmeduranki's parallel to 'Enoch' would lean to the Cainite one, whose distinction was the establishment of a new city. In the Sumerian texts, the events concerning Enmeduranki no longer take place in Eridu, but rather in a new center called Sippar, where he reigned for 21,600 years.

The discovered texts relate how the gods Shamash and Adad took Enmeduranki to the celestial Assembly of the gods, where the secrets of medicine, astronomy, mathematics, etc., were revealed to him. He was then returned to Sippar so that he could start a line of priest-savants:

> Enmeduranki was a prince in Sippar,
> beloved of Anu, Enlil, and Ea.
> Shamash, in *E.babbar,* the Bright Temple,
> appointed him as priest.
> Shamash and Adad [took him]
> to the Assembly [of the gods].
>
> Shamash and Adad clothed (purified?) him.
> Shamash and Adad set him up
> on a large throne of gold.
> They showed him how to observe oil on water—
> a secret of Anu, Enlil, and Ea.
>
> They gave him a Divine Tablet,
> the *Kibdu,* a secret of Heaven and Earth.
> They put in his hand a cedar instrument,
> a favorite of the great gods.
> They taught him how to make
> calculations with numbers.

The two gods, Shamash and Adad—a grandson and a son, respectively, of Enlil—then returned Enmeduranki to Sippar, instructing him to report his divine encounter to the populace and to make his acquired knowledge available to humankind—knowledge that shall be passed from generation to generation, from father to son, by a priestly line beginning with him:

> The learned savant,
> who guards the secret of the great gods,
> will bind his favored son with an oath
> before Shamash and Adad.
> By the divine tablets, with a stylus,
> he will instruct him in the secrets of the gods.

"Thus," states the postscript in the tablet, "was the line of priests created—those who are allowed to approach Shamash and Adad."

In this Sumerian version of the Enoch tale, the two gods acted as the two archangels in the *Book of Enoch* version; it was a theme common in Mesopotamian art, in which two 'Eaglemen' were depicted flanking a gateway (see Fig. 58), a Tree of Life, or a rocket (Fig. 72).

Though in the legible parts of the Enmeduranki tablets his demigodness is not asserted beyond the statement that he "was a Prince in Sippar," his inclusion in the list of pre-Diluvial rulers with a reign of six *Shars* (= 21,600 Earth-years) should serve as an indicator: No mere mortal Earthling could have lived that long. At the same time, such longevity was far short of that of the real Anunnaki gods; Enki, for example, lived through the full 120 Shars from Arrival to Deluge—and he was already an adult when arriving, and stayed on Earth beyond the Deluge. If the eight who reigned after Alulim and Alalgar were not full fledged gods, *they must be considered to have been demigods.*

How can this conclusion be reconciled, say, in the case of the tenth ruler, the hero of the Deluge, if the Bible (re. Noah) lists him as a son of Lamech, and the Sumerian texts (re. Ziusudra) as a son of Ubar-Tutu? The explanation lies in demigod tales, from Bath-Enosh (the

Figure 72

mother of Noah) all the way to Olympias (the mother of Alexander):

Assuming the identity of the husband, a god did it!

Such an explanation admirably affirms the child's demigod status while it absolves the mother of adultery.

An interesting example that illustrates the universality of this explanation comes to us from Egypt, where some of the best known Pharaohs bore theophoric names with the suffix *MSS* (also rendered *MES*, *MSES*, *MOSÉ*) that meant 'Issue/offspring of'—such as Thoth*mes* ('Issue of the god Thoth'), Ra*mses* ('Issue of the god Ra'), etc.

A case in point occurred when Egypt's famed 18th dynasty Pharaoh Thothmes I died in 1512 B.C. He left behind a daughter (Hatshepsut) mothered by his legitimate spouse, and a son mothered by a concubine. Seeking to legitimize his assumption of the throne, the son (thereafter known as Thothmes II) married his half-sister Hatshepsut. The marriage produced only daughters; and when Thothmes II died (in 1504 B.C.) after a short reign, the only male heir was a son not by Hatshepsut, but by a harem girl.

Since the boy was too young to rule, Hatshepsut was appointed co-regent with him. But then she decided that kingship was rightfully hers alone and assumed the throne as a full-fledged Pharaoh in her own right. To justify and legitimize that, she came out with a claim that while Thothmes I was her nominal father, she was actually conceived when the god Amon (= 'The Unseen Ra')—*disguising himself as the husband-king*—was intimate with her mother.

On Hatshepsut's orders, the following statement was included in Egypt's royal annals to record her demigod origins:

> The god Amon took the form of his majesty the king,
> the husband of this queen.
> Then he went to her immediately,
> and he had intercourse with her.
>
> These are the words which the god Amon,
> Lord of the Thrones of the Two Lands,
> spoke thereafter in her presence:

> 'Hatshepsut-by-Amon-created'
> shall be the name of this daughter of mine
> whom I have planted in your body . . .
> She will exercise beneficial kingship
> in this entire land.

Hatshepsut died as Queen of Egypt in 1482 B.C., whereupon the 'boy'—thereafter known as Thothmes III—finally became Pharaoh. Her great and magnificent funerary temple at Deir-el-Bahari, on the Nile's western side opposite ancient Thebes (today's Luxor-Karnak), still stands; and on its inner walls, the story of Hatshepsut's demigod birth is told in a series of murals accompanied by hieroglyphic writing.

The murals start with a depiction of the god Amon, led by the god Thoth, entering the nighttime chamber of queen Ahmose, wife of Thothmes I. The accompanying hieroglyphic inscriptions explain that the god Amon was disguised as the queen's husband:

> Then entered the glorious god, Amon himself,
> Lord of the thrones of the Two Lands,
> having taken the form of her husband.

"They (the two gods) found her (the queen) sleeping in the beautiful sanctuary. She awoke at the perfume of the god [and] merrily laughed in the face of his majesty." As Thoth discreetly left, Amon—

> Enflamed with love, hastened toward her.
> She could behold him, in the shape of a god,
> as he came nearer to her.
> She exulted at the sight of his beauty.

Both enamored, god and queen had inercourse:

> His love entered into all her limbs.
> The place was filled with the god's sweet perfume.
> The majestic god did to her all that he wished.
> She gladdened him with all of herself;
> she kissed him.

Attributions of liaisons by Ra that endowed future Egyptian Pharaohs with demigod status go back, in fact, to earlier dynastic times. A tale, inscribed on papyrus may even solve a mystery concerning Egypt's 5th dynasty in which three related Pharaohs succeeded each other without being fathers-sons. According to that tale, they were conceived when the god Ra mated with the wife of the high priest of his temple. When the pangs of childbirth began, it was realized that the woman carries a triplet and would have a very difficult time giving birth. So Ra summoned four 'birth goddesses' and appealed to his father, Ptah, to assist in the births. The text describes how all those gods assisted as the wife of the high priest gave birth, in succession, to three sons who were named Userkaf, Sahura, and Kakai. Historical records show that the three of them indeed reigned in succession as Pharaohs, forming the Fifth Dynasty; they were a Demigod Triplet.

Besides providing Egyptologists with an explanation of that odd dynasty, the tale also offers an explanation for a bas-relief, discovered by archaeologists, that depicts the Pharaoh Sahura as a baby suckled by a goddess—a privilege limited to those of divine birth. Such 'divine suckling' was also claimed by Hatshepsut to further her claim to divinely ordained kingship: she asserted that the goddess Hathor (nicknamed 'Mother of gods') suckled her. (A successor, the son of Thothmes III, also claimed to having been divinely suckled.)

A claim of direct demigod status as a result of intercourse with a god in disguise was then made by the famed Ramses II by recording in the royal annals the following revelation that the great god Ptah himself made to the Pharaoh:

> I am thy father.
> I assumed my form as *Mendes,* the Ram Lord,
> and begot thee inside thy august mother.

If such a claim to having been fathered by not just one of the gods but by the head of the pantheon looks too far fetched, recall our explanation that the god called **Ptah** by the Egyptians was none other than **Enki.**

And to assert a fathering by Enki was not outlandish at all.

* * *

As one takes a sweeping view of the Mesopotamian tales of the gods, there come into focus the different personalities of the half-brothers Enki and Enlil—in every respect, including matters of sexual behavior.

Anu, we have earlier mentioned, had quite a harem of concubines in addition to his official spouse, Antu; indeed, the mother of Ea/Enki, Anu's firstborn son, was one such concubine. When Anu and Antu came to Earth on a state visit (circa 4000 B.C.), a special city, Uruk (the biblical Erech), was built to accommodate them. During the visit Anu took a special liking to Enlil's granddaughter, who was called thereafter **In.Anna** (= 'Anu's Beloved')—with hints, in the texts, that Anu's "loving" was not just grandfatherly.

And in these respects, Enki and definitely not Enlil, had his father's genes. Of his six sons, only Marduk is clearly identified as mothered by Enki's official spouse **Dam.ki.na** (= 'Lady [who] to Earth Came'); the other five sons' mothers are mostly unnamed and could have been concubines or (see hereunder) chance encounters. By comparison, Enlil—who had a son by Ninmah̲ back on Nibiru when both were unmarried—had sons (two) only by his spouse, **Ninlil**.

A long Sumerian text that its first translator, Samuel N. Kramer, named *Enki and Ninh̲ursag: A Paradise Myth,* details Enki's repeated sexual intercourses with his half-sister Ninh̲arsag/Ninmah̲ in (unsuccessful) attempts to have a son by her, and then his intercourses with the female offsprings of those liaisons. (Ninh̲arsag—a medical officer—had to inflict Enki with painful maladies to make him stop.) As often as not, these Enki tales extolled the god's mighty penis.

Enki was not averse to keeping sex within the family: A long text dealing with Inanna's visit to Eridu (to obtain from Enki the vital Mé) describes how her host attempted (unsuccessfully) to get her drunk and seduce her; and another text, recording a voyage from Eridu to the Abzu, relates how Enki did succeed to have sex with **Ereshkigal** (Inanna's elder sister and future wife of Enki's son **Nergal**) aboard their boat.

When such escapades resulted in the birth of offspring, young gods or goddesses were born; for *demigods* to be born, the intercourse had to be with Earthlings; and of that too there was no shortage . . . We can begin with Canaanite tales of the gods, where ***El*** (= 'The Lofty One'—'Cronos' of eastern Mediterranean lore) was head of the pantheon. The tales include a text known as *The Birth of the Gracious Gods;* it describes how El, strolling on the seashore, met two Earthling females bathing. The two women were charmed by the size of his penis and had intercourse with him, resulting in the birth of ***Shaẖar*** (= 'Dawn') and ***Shalem*** (= 'Complete' or 'Dusk').

Though called 'gods' in the Canaanite text, the two were, by definition, demigods. An important epithet-title of El was *Ab Adam*—translated 'Father of Man' but also meaning ***'Father of Adam,'*** which, literally taken, may mean just that: Progenitor and actual father of the individual the Bible calls Adam, as distinct from the prior references to "The Adam" species. And this leads us directly to ***Adapa,*** the legendary Model Man of Mesopotamian texts.

A pre-Diluvial demigod known as "Man of Eridu," his name, *Adapa,* identified him as the "Wisest of Men." Tall and big of size, he was most clearly identified as a son of Enki—a son of whom Enki was openly proud, whom he appointed as Chief of Household in Eridu, and to whom he granted "wide understanding"—all manner of knowledge, including mathematics, writing, and craftsmanship.

The first "Wise Man" on record, Adapa might have been the elusive *Homo sapiens sapiens* who appeared on the human scene some 35,000 years ago as 'Cro-Magnon Man', as distinct from the cruder Neanderthals. It has been speculated (with no convincing conclusion) whether 'Adapa' could have been the actual person the Bible calls 'Adam' (as distinct from 'The Adam' species). I, for one, wonder whether he could have been the ***En.me.lu.anna*** of the pre-Diluvial Sumerian King Lists—a name translatable as 'Enki's Man of Heaven'—for the most memorable and unique event concerning Adapa was ***his celestial journey to visit Anu on Nibiru.***

The Tale of Adapa begins by giving the reader the sense of a very long ago time, at the beginning of events, when Ea/Enki was involved in Creation:

In those days, in those years,
by Ea was the Wise One of Eridu
created as a model of Man.

The tale of Adapa reverberated in Mesopotamian life and literature for ages. Even in later Babylon and Assyria, the expression "Wise as Adapa" was used to describe someone highly intelligent. But so was another aspect of the Adapa tale, according to which Ea/Enki deliberately granted one but withheld another divine attribute from this Model of Man, though his own son:

Wide understanding he perfected for him;
Wisdom he had given him;
To him he had given Knowledge—
Everlasting life he had not given him.

As word reached Nibiru of the unusually wise Earthling, Anu asked to see Adapa. Complying, Enki "made Adapa take the way to Anu, and to heaven he went up." But Enki was concerned lest Adapa, while on Nibiru, be offered the Bread of Life and the Water of Life—and attain the longevity of the Anunnaki after all. To prevent that from happening, Enki made Adapa look wild and shaggy, dressed him shabbily, and gave him misleading instructions:

As you stand before Anu,
they will offer you bread;
it is Death—do not eat!
They will offer you water;
it is Death—do not drink!
They will offer you a garment—
put it on.
They will offer you oil—
anoint yourself with it.

"You must not neglect these instructions," Enki cautioned Adapa; "to that which I have spoken, hold fast!"

Taken aloft by "the Way of Heaven," Adapa reached the Gate of Anu; it was guarded by the gods Dumuzi and Gizidda. Let in, he was brought before Anu. As Enki had predicted, he was offered the Bread of Life—but fearing death, refused to eat it. He was offered the Water of Life, and refused to drink it; he did put on the clothes he was offered, and anointed himself with the oil given him. Puzzled and bemused, Anu asked him: "Come now, Adapa—why did you not eat, why did you not drink?" To which Adapa answered: "Ea, my master, commanded me 'you shall not eat, you shall not drink'."

Angered by the answer, Anu sent an emissary to Enki, demanding an explanation. The inscribed tablet is too damaged here to be legible, so we don't know Enki's response. But the tablet does make it clear that Adapa, having been found "worthless" by Anu, was returned to Earth and started a line of priests adept at curing diseases. Wise and intelligent, a son of the god Enki Adapa was—yet as a mortal he died.

The scholarly debate whether the biblical 'Adam' was 'Adapa' is yet to be settled. But clearly, the biblical narrator had the tale of Adapa in mind when writing the story of the two trees in the Garden of Eden—the Tree of Knowing (of which Adam ate) and the Tree of Life (of which he was precluded). The warning to Adam (and Eve), "the day you shall eat thereof surely you shall die," is almost a quote from Enki's warning to Adapa. So is the deity's concern, expressed to unnamed colleagues, regarding the risk of Adam eating also from the Tree of Life (Genesis 3:22–24):

> And *Yahweh Elohim* said:
> Behold, the Adam is become as one of us
> to know good and evil;
> And now, what if he put forth his hand
> and took also of the Tree of Life,
> and ate, and lived forever?

So "*Yahweh Elohim* expelled him from the Garden of Eden . . . and placed at the east of the Garden of Eden the *Cherubim,* and the flaming sword which revolveth, to guard the way to the Tree of Life."

We do not know whether Enki's warning to Adapa—to avoid the Water and Bread of Life lest he dies—was an honest one, or part of the deliberate decision to give Adapa Wisdom but not "Everlasting life." We do know, however, that the warning to Adam and Eve, that they will "surely die" were they to eat of the Tree of Knowing, was untrue. God, as the Serpent told them, lied.

It is an episode to be kept in mind as the issues of Immortality will come to the fore.

* * *

According to the WB-62 king list, Enmeluanna was followed by En.sipa.zi.anna (= 'Shepherd Lord, Heavenly Life') and then by Enmeduranna/Enmeduranki, whose tale matches that of the biblical Enoch. Different and ambiguous names are then given by the Mesopotamian sources for the biblical Lamech, the most certain of which is the ***Ubar-Tutu*** in the Epic of Gilgamesh (and thus probably the *Obart*es of Berossus). Nothing, apart from this mention in the Epic of Gilgamesh, is known about that predecessor of Ziusudra/Utnapishtim. Was he a demigod, or the hapless Lamech who had doubts about the true parentage of Noah?

The 'transgressions' by the Igigi or "Watchers" that so upset Enlil were in fact begun by none other than Enki himself. They resulted, as the varied sources make clear, in numerous demigod offspring; but only a handful of them are named and listed. Were they the instances in which Enki himself, bearing the epithet En.me, was involved?

The riddle of Patriarch-Demigods in pre-Diluvial times runs all the way to Noah and the Deluge; but the enigma of our ancestral "seed" does not end there, for—as the Bible states (and Mesopotamian sources confirm)—the intermarriage that began before the Deluge continued "also after that."

We will soon find that other gods—and goddesses!—were eager intermarriage partners in the post-Deluge times.

WORDS AND THEIR MEANING

Readers of transliterated Sumerian texts may find a small 'd' prefixing a deity's name—e.g., dEnki, dEnlil. Called a 'determinative', it identifies the name as that of a god (or goddess). The d is shorthand for the two-syllable word ***Din.gir***. Literally meaning 'Righteous Ones [of the] Rocketships', it was depicted pictographically as a rocket with a command module (see sidebar "The Land of 'Eden'" on page 82). Simplified, the designation 'god/divine' was rendered by a 'star' sign that was read ***An*** and evolved further to a crosslike wedgemark (see illustration); it was read *Ilu* in Akkadian (i.e., Babylonian, Assyrian)—from which the singular *El* in Canaanite or Hebrew and the plural *Elohim* in the Bible.

While explaining that in the tale of Adam's creation, etc., the *Elohim* of the Bible were the Sumerian Anunnaki, this author (as unambiguously stated in *Divine Encounters*) envisions God (with a capital 'G') as a universal cosmic Creator of All, who acts through emissaries—'gods' with a small 'g'. *The existence of the Elohim/Anunnaki 'gods' with a small 'g' is confirmation of the existence of* **their creator**, *God with a capital 'G'.*

The encompassing divine name "***Yahweh***" was explained to Moses as meaning *Eheyeh asher eheyeh*—"I will be whoever I will be"—God could "be" (act through) Enki in one instance, or "be" through Enlil in another instance, etc. When the Hebrew text states *Elohim,* Anunnaki 'gods' are spoken of; and when the Bible employs the term *Yahweh Elohim,* it should be recognized as meaning 'When Yahweh acted as/through one of the *Elohim*'.

Other unorthodox understandings of biblical words suggested in my writings, include the term ***Olam***. It is commonly translated 'Forever/Everlasting/of old'; but stemming from the root verb that means "To hide," *Olam* (I wrote) could mean a physical 'Hidden Place' of God, as in Psalm 93:2—"Thou art ***from*** *Olam*"—the 'Hidden Place,' the unseen planet Nibiru.

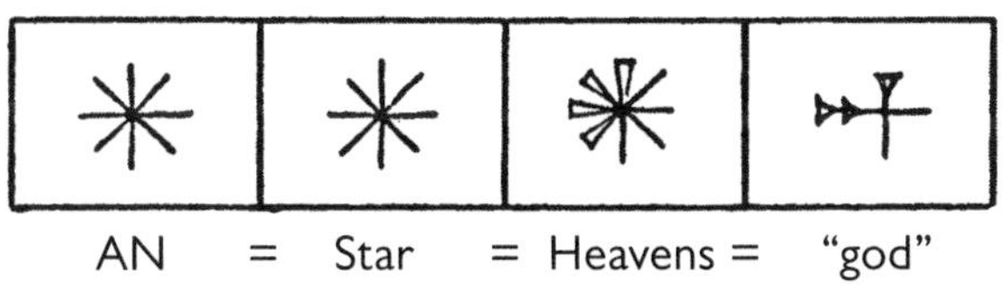

AN = Star = Heavens = "god"

XI
There Were Giants Upon the Earth

There were *giants* upon the Earth
in those days ***and thereafter too.***

With a few (by now familiar) words—highlighted above—the Bible extended the pre-Diluvial epic events involving the demigods, to post-Diluvial days; one can even say, from prehistoric and legendary ages to historical times.

The reader knows by now that Genesis verse 6:4 does not say 'giants'—it says *Nefilim,* and that I was the schoolboy who questioned the teacher on his explanation of 'giants' rather than the 'Those who have come down' meaning. In retrospect, I realized that the teacher did not invent the 'giants' interpretation, and that there had to be a reason why the scholars assigned by King James I of England to translate the Hebrew Bible used the term 'giants': They relied on earlier translations of the Hebrew Bible—in Latin known as the *Vulgate,* dating back to the 4th and 6th centuries A.D. and a prior Greek translation (the *Septuagint*) done in Alexandria, Egypt, in the 3rd century B.C. And in both those early translations, the word *Nefilim* is rendered "gigantes." Why?

The answer is given in the Bible itself. The term *Nefilim,* first employed in Genesis 6:4, is used again in the Book of Numbers (13:33), in the tale of the scouts that Moses sent ahead to scout Canaan as the Israelites readied to enter it at the end of the Exodus. Selecting twelve men, one from each tribe, Moses told them: "Go up from the Negev (the southern dry plain) unto the hills, and see the country—what is it like, and who are the people that dwell therein—are they strong or are they weak? Many or a few? And what is the land in which they dwell, is it good or bad? And what are the cities that they inhabit—are they in open fields, or are they fortified?"

Proceeding as instructed, the twelve scouts, "Coming up from the Negev, reached Hebron, where Achiman, Sheshai, and Talmai, the descendants of *Anak,* were." And when the scouts returned, they said to Moses:

> We came unto the land whither thou didst send us,
> and truly doth it flow with milk and honey . . .
> But the people who dwell in the land are strong,
> and the cities are large and fortified;
> and we even saw there the children of *Anak.*
>
> We did see there ***the giants,***
> the ***sons of Anak—the Nefilim,***
> the children of Anak of the *Nefilim;*
> and we were like grasshoppers in our eyes,
> and so were we in their eyes.

The singular *Anak* is also rendered in the plural, ***Anakim,*** in Deuterenomy 1:28 and 9:2, when Moses encouraged the Israelites not to lose heart because of those fearsome "descendants of Anak"; and again in Joshua 11 and 14, in which the capture of Hebron, the stronghold of the "Children of the Anakim," was recorded.

As those verses equate the Nefilim with the Anakim, they also depict the latter (and thus the former) as giantlike—so big that average Israelites were like grasshoppers in their eyes. Capturing their fortified strongholds, with particular attention to Hebron, was a special achievement in the Israelite advance. When the fighting was over, the Bible states, "There remained not *Anakim* in the land of the Children of Israel except those who were left over in Ghaza, Gath, and Ashdod" (Joshua 11:23). The uncaptured strongholds were all cities of a Philistine coastal enclave; and therein lie additional reasons for equating the Anakim with giants—for King David's giantlike Philistine adversary *Golyat* ('Goliath' in English) and his brothers were descendants of the Anakim who remained in the Philistine city of Gath. According to the Bible, Goliath was more than nine feet tall; his name became a synonym for 'giant' in Hebrew.

The name *Gol-yat,* of unknown origin, may well contain a hitherto unnoticed connection to the Sumerian language, in which ***Gal*** meant 'Large/big/great'—as discussed in greater detail in ensuing paragraphs.

It was only after concluding that the biblical *Nefilim* were the *Anunnaki* of Mesopotamian lore that it dawned on me that *Anakim* was simply a Hebrew rendering of the Sumerian/Akkadian *Anunnaki.* If this original insight yet simple equation has not yet been universally adopted, the reason can only be the established view that whereas the Anakim as sons of the Hebronite Anak could have existed, the Anunnaki gods—don't we all know?—were just a myth . . .

The Anakim-Anunnaki connection finds additional corroboration in an unusual choice of terminology in Joshua 14:15. Describing the capture of Hebron as the feat that brought the fighting in Canaan to an end, the Bible had this to say about the city (per the King James translation): "And the name of Hebron before was Kiriath Arba, which Arba was a *great man* among the Anakim."

More modern English translations of this statement offer some variations regarding the identity of ***Arba***. The New English Bible renders it "Formerly the name of Hebron was Kiriath-Arba; this Arba was the *chief man* among the Anakim." The New American Bible translates "Hebron was formerly called Kiriath-arba, for Arba, *the greatest* among the Anakim." And the new *Tanach* Jewish Bible says, "The name of Hebron was formerly Kiriath-arba, [Arba] was the *great man* among the Anakites."

The translation problem stems from the fact that the Hebrew text describes Arba as "the *Ish Gadol* of the Anakim." Literally translated, *Ish* unambiguously means a male Man; but *Gadol* could mean both 'Big/Large' as well as 'Great'. So, was the intention of this descriptive epithet to say that Arba was a *Big* Man in size—a 'Goliath'—or a *Great* Man—an outstanding leader?

As I was reading and re-reading this verse, it struck me that I have come across this exact term—*Ish Gadol*—before: ***In the Sumerian texts!*** For in them, the term that denoted 'king' was ***Lu.gal***—literally ***Lu*** (= 'Man') + ***Gal*** (= 'Big/Great') = *Ish-Gadol*. And, as in the Hebrew, ***the term had its ambiguous double meaning: Big/Large Man or 'King' (= 'Great Man').***

And here another thought occurred: Was there perhaps no ambiguity—was this 'Arba', descendant of the Anunnaki, a demigod who ***was both large/big and great?***

The pictograph from which the cuneiform signs for ***Lugal*** evolved showed the symbol for ***Lu*** to which a crown was added (Fig. 73), and it does not indicate size. We don't have a picture of Arba (whose name literally meant 'He who is Four'); but we do have ancient depictions of Sumerian kings; and in the Early Dynastic period they were depicted as

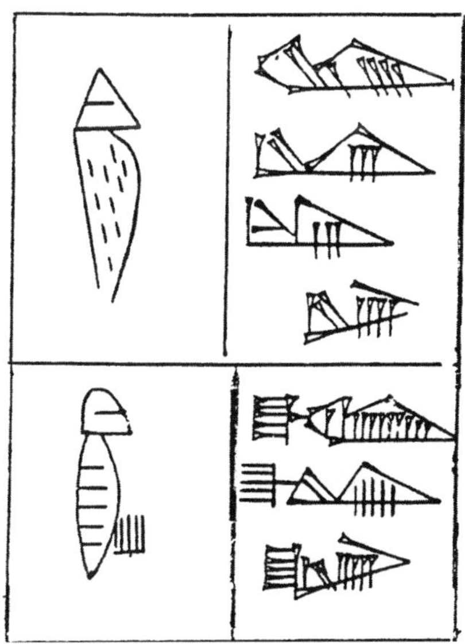

Figure 73

Figure 74

big fellows (for example, Fig. 74). Other examples from Ur, circa 2600 B.C., are the depictions on a wooden box known as 'The Standard of Ur' with panels on its two sides, one (the 'War Panel', Fig. 75) showing a scene of marching soldiers and horse-drawn chariots, and the other (the 'Peace Panel') of civilian activities and banqueting; the person who stands out by his big size is the king—the ***Lu.gal*** (Fig. 76, portion of panel).

(It might be relevant to mention here that when the Israelites decided to have a king, the one chosen—Saul—was picked because "when he stood among the people, he was taller than any of the people from his shoulder and upward." I Samuel 10:23.)

Of course, not all kings in antiquity were giantlike. A Cannanite Big One, Og the King of Bashan, was so unusual that the Bible makes a point of it. Arba—descended from the Anakim/Anunnaki—stood out because he was *Ish Gadol*. Though not a king, the demigod Adapa—son of Enki—was described as big and robust. If such 'Big

Figure 75

Figure 76

Man' demigods inherited that genetic trait from their divine parents, one would expect depictions of gods and men to also show the deities as relatively giantlike; *and that was actually the case.*

Figure 77

Figure 78

It can be seen, for example, in a 3rd millennium depiction from Ur of a naked ***Lugal,*** bigger than the people bearing offerings behind him, pouring a libation to an even bigger seated goddess, Fig. 77. Similar depictions have been found in Elam; and the same 'ratio' of king-to-deity is also seen in a depiction of a big Hittite king offering a libation to an even bigger god Teshub (Fig. 78). Another perspective of this theme can be seen in Fig. 51, in which a lesser deity introduces a king to a seated god who—were he to stand up—would be at least one-third taller than the others.

Such bigness, one finds, was not limited to male gods; **Ninmah̲/ Ninh̲arsag** (who in her old age was nicknamed 'The Cow') was depicted as hefty (Fig. 79). More famous for her size, even in her younger days,

Figure 79

Figure 80

was the goddess **Ba'u** (Fig. 80), the spouse of the god Ninurta; her epithet was ***Gula*** (= 'The Big One').

There were, indeed, giants upon the Earth in pre-Deluge times, and thereafter too. Luckily, the great archaeological discoveries of the past two centuries enable us to identify them and to bring them to life—even as they died.

* * *

In spite of its statement that the *Gibborim*—Heroes, *'Mighty Men'* (alias demigods)—continued into post-Diluvial times, the Bible makes hardly any mention of them until the Israelite return to Canaan. It was only then, when Moses recounted who had inhabited Canaan, that the Bible mentions the ***Anakim*** and a sub-group called ***Repha'im*** (a term that might mean 'Healers') who per Deuteronomy 2:11 "as *Anakim* are considered." Most (except for certain 'Children of Anak') were replaced by a variety of tribe-nations who repopulated those lands after the Deluge.

According to the Bible, it was of Noah's three sons—***Shem, Ham,*** and ***Japhet***—who had survived the Deluge with their wives, that Mankind re-emerged: "It is of them that the whole Earth was overspread," the Bible stated as it launched a list of their descendant-nations (Genesis, chapter 10). And in that long and comprehensive list, only one sole heroic figure called ***Nimrod*** is named.

Stemming from Kish (misspelled 'Kush'), Nimrod "was a *Mighty Hunter* by the grace of Yahweh"; it was he "who was the *First Hero* in the land" per the Genesis verses that we have already quoted. We mentioned earlier the scholarly assumption, upon the discovery and decipherment of the cuneiform tablets, that 'Nimrod' (whose domains included Erech in the land of Shine'ar) was the famed Sumerian Gilgamesh, king of Erech/Uruk—an incorrect assumption, as it turned out. But the Hebrew epithets applied to Nimrod—a *Gibbor,* a Hero, a *Mighty* hunter—unmistakably links him to the plural *Gibborim* of Genesis 6:4, and thus identifies him as one of the continued line of demigods. (In Sumerian iconography, it was **Enlil** who was depicted as the granter of a hunting bow to Mankind, Fig. 81).

Figure 81

The assertion that Nimrod was "brought forth" in *Kish* can serve as an invaluable clue regarding his identity; it lurks, I believe, ***unrecognized among the demigods associated with the god Ninurta***. It certainly links these biblical verses to the Sumerian King List, where it is stated in regard to the post-Diluvial period:

> After the Flood had swept thereover,
> when kingship was lowered (again) from heaven,
> *the kingship was in Kish.*

Kish was not one of the pre-Diluvial cities that were rebuilt exactly where they had been once Mesopotamia was habitable again; it was a new city, intended as a neutral capital, whose establishment followed the creation of separate regions for the contending Anunnaki clans.

The Deluge calamity that befell the Earth—a colossal tidal wave caused by the collapse of the ice sheet over Antarctica—unavoidably overwhelmed the Abzu with its gold-mining facilities in southeastern Africa. But as nature would have it, the calamity that destroyed one side of the Earth had beneficial effects on the other side: In the Lands Beyond The Seas that we now call South America, the powerful avalanche of water exposed extremely rich veins of gold in the (now called) Andes mountains, and filled riverbeds with easily collected gold nuggets. As a result, the gold that Nibiru needed could be obtained there without the toil of mining. Preempting Enki, Enlil sent his son Ishkur/Adad to take charge of the golden territory. Control of the repopulated olden lands thus became a pressing issue for the 'deprived' Enki's clan; the suggested creation of distinct regions and clearly delineated territories was an attempt at peacemaking by Ninmaḫ.

Before Kingship was reinstated on Earth after the Deluge, a text dealing with the matter states, "The great Anunnaki gods, the deciders of destinies, sat in council, made decisions concerning the Earth, and established the four regions." The allocation of three regions matched the three biblical nation-state branches emanating from the three sons of Noah; its purpose and result was to allot Africa (and the

Hamitic peoples) to Enki and his sons, and Asia and Europe (Semitic and Indo-European peoples) to Enlil and his sons. A Fourth Region, territory of the gods alone, was set aside for a new, post-Diluvial Spaceport; located in the Sinai peninsula, it was placed under the aegis of the neutral Ninmah, earning her the epithet **Nin.harsag** (= 'Lady/Mistress of the Mountain peak'). Called ***Til.mun*** (= 'Place/Land of the Missiles'), it was the place to which Ziusudra and his wife were taken after the Deluge.

The principal aim of forming the regions—a 'share and share alike' arrangement between and within the Anunnaki clans—was not readily attained. Discord and strife soon broke out among the Enki'ites; Egyptian lore recalled it as first the struggle for dominion between Seth and Osiris, leading to the killing of Osiris, then in revenge warfare between Horus (born of the semen of Osiris) and Seth. Enki's son Marduk (**Ra** in Egypt) repeatedly tried to establish himself in Enlilite territories. A relatively peaceful era—negotiated by Ninmah—was again shattered by a rivalry between Enki's sons Ra/Marduk and Thoth/Ningishzidda. It took another millennium to restore Earth and Mankind to stability and prosperity, making possible Anu's state visit to Earth, circa 4000 B.C.

The Bible asserts that the fourth-generation descendant of Shem was named *Peleg* (= 'Division'), "because in his time was the Earth divided"; in *The Wars of Gods and Men* I have suggested that this was a reference to the establishment of the three separate regions of civilization—of the Euphrates/Tigris, the Nile, and the Indus Rivers. Peleg was born, according to the Bible, 110 years after the Deluge; using the 'times sixty' formula, it would date the birth of Peleg to circa 4300 B.C. (10,900–6,600) and the "division" to circa 4000 B.C.

With the creation of Mankind's civilizations, Enlil's post-Diluvial headquarters in ***Ni.ibru*** (*Nippur* in Akkadian)—established precisely where the pre-Diluvial city had been, but no longer Mission Control Center—became the overall religious capital, a kind of 'Vatican'. It was then that a luni-solar calendar, the *Calendar of Nippur,* with a cycle of twelve ***Ezen*** (= 'Festival') periods—the origin of 'months'—was fixed. That calendar, begun in 3760 B.C., is still followed as the Jewish Calendar to this day.

And then the gods "mapped out the city of Kish, laid out its foundations." It was intended as a national capital, a kind of 'Washington D.C.'; and it was there that the Anunnaki started the line of post-Diluvial kings by "bringing down from heaven the scepter and crown of kingship."

* * *

The excavations conducted at the site of ancient Kish, described in our chapter 4, have corroborated varied Sumerian texts that named the god Ninurta as that city's titular deity, giving rise to the thought that perhaps he was the 'Nimrod' who was Yahweh's "Mighty Hunter." But the Sumerian King List actually named the first ruler in Kish; regrettably we still don't know it, because the inscription is damaged right there, leaving legible only the syllables ***Ga.--.--.ur***. What is clearly legible is the statement that he reigned for 1,200 years!

The name of the second ruler in Kish is entirely damaged, but his reign lasted a clearly written 860 years. He was followed on the throne of Kish by ten legibly named kings with reigns lasting 900, 840, 720, and 600 years. Since these are numbers clearly divisible by 6 or 60, the unanswered question is whether these are factual reign lengths, or did the ancient copying scribes misread them, and it should have been 200 (or 20) for Ga.--.--.ur, 15 instead of 900 for the next one, etc. Which was it?

The 1,200 year reign of Ga.--.--.ur, if correct, places him in the category of the pre-Diluvial biblical Patriarchs (who lived almost 1,000 years each), and his immediate successors somewhat ahead of Noah's sons (Shem lived to 600). If Ga.--.--.ur was a demigod *Gibbor,* 1,200 years in his case might be plausible. So would be the 1,560 years attributed to the 13th king in Kish, ***Etana,*** regarding whom the King List makes the long notation: "A shepherd, he who to heaven ascended, who consolidated the countries." In this case, the royal notation is supported by discovered literature, including an ancient two-tablet text relating

The Etana Legend, for he was indeed a king who "to heaven ascended."

A benevolent ruler, Etana was despondent by the lack of a male heir, caused by his wife's pregnancy difficulties that could be cured only by the heavenly Plant of Birth. So he appealed to his patron god Utu/Shamash to help him obtain it. Shamash directed him to an "eagle's pit"; and after overcoming varied difficulties the Eagle took Etana aloft to the "Gate of Anu's heaven."

As they rose ever higher, the Earth below them appeared ever smaller:

> When he had borne Etana aloft one *beru,*
> the Eagle says to him, to Etana:
> "See, my friend, how the land appears!
> Peer at the sea at the side of the mountain house—
> The land has become a mere hill,
> the wide sea is just like a tub."

Rising a second *beru* (a measure of distance as well as degrees of the celestial arc), the Eagle again urged Etana to look down:

> "My friend,
> Cast a glance at how the Earth appears!
> The land has turned into a furrow . . .
> The wide sea is just like a bread basket!"

"After the Eagle had carried him aloft a third beru," the land "turned into a gardener's ditch." And then, as they continued to ascend, the Earth suddenly disappeared from view; and—as the frightened Etana later said: "As I glanced around, the land had disappeared!"

According to one version of the tale, Etana and the Eagle "passed through the gate of Anu." According to another version, Etana became alarmed and cried out to the Eagle: "I am looking for the Earth, but I cannot see it!" Frightened, he shouted to the Eagle: "I cannot go on to the heavens! Take the road back!"

Heeding the cries of Etana, who was "laying slumped on the Eagle's wings," the Eagle fell back to Earth; but (according to this version), Etana and the Eagle made a second attempt. It was apparently successful, for the next king in Kish, Baliẖ, is identified as "son of Etana." He reigned a mere 400 (or 410) years.

The tale of Etana was depicted by ancient artists on cylinder seals (Fig. 82), one that starts with the 'Eagle' in its 'pit', and another that shows Etana hovering between the Earth (= 7 dots) and the Moon (identified by its crescent). The tale is instructive in several respects: It describes realistically a flight out to space with a diminishing Earth in sight; it also corroborates what many other texts suggest—that comings and goings between Earth and Nibiru were more frequent than once in 3,600 years. The tale does leave Etana's mortal vs. demigod status unstated; but one can only surmise that Etana would not have been allowed the space flights, nor would have reigned a purported millennium and a half, were he not a demigod.

The fact that a later inscription prefixes Etana's name with the 'Dingir' determinative reinforces a conclusion that Etana indeed was divinely engendered; and a notation in another text that Etana was of

Figure 82

the same "Pure Seed" of which Adapa had been, can serve as a clue to who the father was.

The possibility that the 23 kings who reigned in Kish alternated between demigods and their mortal offspring comes especially to mind as we reach the 16th king, ***En.me.nunna,*** who ruled for 1,200 years and was followed by his two sons with mortal-like reigns of 140 and 305 years. There followed kings reigning 900 and 1,200 years; and then ***En.me.bara.ge.si,*** "who carried away as spoil the weapons of Elam, became king and ruled 900 years."

Though the *Shar* counts are gone, the two theophoric names sound familiar; they place these post-Diluvial kings in the same names category as the pre-Diluvial ones (of the WB tablets and the Berossus list) *who had gods as parents.* They also provide a historical dimension to the Kish list, for the name Enmenbaragesi was found inscribed on an archaeological artifact—a stone vase now in the Iraq Museum in Baghdad; *Elam* (whose weapons he took as spoils) was an historically verified kingdom.

Aka, son of Enmebaragesi, who reigned for 629 years, completed the list of 23 kings of Kish who "reigned a total of 24,510 years, 3 months, and three and a half days"—some four millennia if divided by 6, only four centuries if reduced by 60. And then Kingship in Sumer was transferred to Uruk.

* * *

The seat of central Kingship was transferred from Kish to Uruk some time circa 3000 B.C.; and right off, we need guess no more who had reigned there, for this is what the King List states about the first king of Uruk:

> In Uruk,
> Mes.kiag.gasher, ***son of dUtu,***
> became high priest and king
> and reigned 324 years.
> Mes.kiag.gasher
> went into the sea
> (and) came out to the mountains.

Though obviously a demigod, fathered by the god Utu/Shamash, no more than 324 years (also a number, please note, divisible by 6) are assigned to him; and no explanation is offered for such a short reign by a full-fledged demigod. His name conveyed the meaning 'Handy, dexterous'. And since no other text about Meskiaggasher has been found, we can only guess that the sea he crossed to reach a mountainland—a voyage that merited its quoted mention—were the Persian Gulf ("Lower Sea") and the land Elam, respectively.

Uruk (the biblical Erech) was established not as a city but as a rest-place for Anu and Antu when they came to Earth for a state visit circa 4000 B.C. When they left, Anu gave it as a gift to his great-granddaughter **Irninni**, nicknamed and better known thereafter as **In.Anna** (= 'Anu's Beloved'), alias **Ishtar**. Ambitious and enterprising—the Great God List records more than one hundred epithets for her!—Inanna, outsmarting the womanizing Enki, managed to obtain from him more than a hundred *Mé* ('Divine Formulas') needed to make Uruk a principal city.

The task of actually reshaping Uruk into major city status was carried out by the next king of Uruk, **Enmerkar**. According to the Sumerian King List, he was "the one who built Uruk." Archaeological evidence suggests that it was he who built the city's first protective walls, and expanded the ***E.Anna*** temple into a sacred precinct worthy of a great goddess, the goddess Inanna. An exquisitely carved alabaster vase from Uruk—one of the most prized objects in the Iraq Museum in Baghdad—depicted a procession of worshippers, led by a giantlike naked king, bringing offerings to the the 'Mistress of Uruk'.

Called in the King List "son of Mes.kiag.gasher," Enmerkar reigned 420 years—almost a century longer than his demigod father. Much more is known of him, for he was the subject of several epic tales, the longest and most historical of which is known as the tale of *Enmerkar And The Lord of Aratta*—one of whose revelations, *most clearly and repeatedly stated,* is that ***Enmerkar's real father was the god Utu/Shamash.*** This made him a direct relative, and not just a worshipper, of Utu's sister Inanna; and therein one finds an explanation for enigmatic journeys to a distant kingdom.

The establishment of Four Regions was intended as a way to restore peace among the Anunnaki clans by a 'let each one have his own' arrangement (the Tigris-Euphrates Plain, under the Enlilites, was the First Region; Africa, under the Enki'ites, was the Second Region). Another idea was to enhance peace through intermarriage; and chosen for the purpose was Enlil's granddaughter Inanna/Ishtar and the shepherd god **Dumuzi**—Enki's youngest son (but only a half-brother of Marduk). References in varied texts suggest that the unassigned Third Region, the Indus River valley, was intended as a dowry for the young couple. (The Fourth Region, from which Mankind was excluded, was the Spaceport in the Sinai Peninsula.)

Arranged marriages were part of the Anunnaki record, both on Nibiru and on Earth; one of the earliest Earth-instances is recorded in a tale of Enki and Nin<u>h</u>arsag: Their lovemaking resulted in the birth of females only, and the two then spent time matching them with spouses. As it happened, the young Inanna and Dumuzi not only liked each other, but fell in love. Engaged to be married, their torrid love and lovemaking is described in long and detailed poems, mostly composed by Inanna, giving her a reputation as Goddess of Love (Fig. 83a). The poems also revealed Inanna's ambition to become, through the marriage, Mistress

Figure 83

of Egypt, and this alarmed Enki's son Marduk/Ra; his efforts to disrupt the marriage led—unintentionally, he claimed—to the death by drowning of Dumuzi.

Lamenting and enraged, Inanna launched fierce battles against Marduk/Ra, establishing her record as a Goddess of War (Fig. 83b). Dubbed by us 'The Pyramid Wars' in *The Wars of Gods and Men,* they lasted several years and ended only with the imprisonment and then exile of Marduk. The great gods tried to console Inanna by granting her sole dominion of the faraway Kingdom of Aratta, situated farther east of Elam/Iran and beyond seven mountain ranges.

In *The Stairway to Heaven* I suggested that the Kingdom of Aratta was the Third Region—what is nowadays described as the Indus Valley Civilization (with its center, called Harappa by archaeologists, on the significant 30th parallel north). Thus, it was the destination of the Meskiaggasher voyage and the locale of the significant events that followed.

The context for the *Enmerkar and the Lord of Aratta* tale was the odd situation that the City of Uruk and the Kingdom of Aratta shared the same goddess, Inanna. Moreover, the unnamed king of Aratta is repeatedly identified as "*seed implanted in the womb by Dumuzi*"—an enigmatic statement that leaves one guessing not only who the mother was, but also whether post-mortem artificial insemination was involved. (An instance of such artificial insemination is recorded in Egyptian tales of the gods, when the god Thoth extracted semen from the phallus of the dead and dismembered Osiris and impregnated with it Isis, the wife of Osiris, who then gave birth to the god Horus.)

Calling himself "Sumer's Junior Enlil," Enmerkar sought to establish superiority for Uruk by refurbishing and enlarging the olden temple of Anu, the E.anna, as Inanna's principal shrine, and by placing Aratta in second-class status by forcing it to send to Uruk 'contributions' of precious stones, lapis lazuli and carnelian, gold and silver, bronze and lead. When Aratta, described in the text as "a highland place of silver and lapis lazuli," delivered the tribute, Enmerkar's heart grew hauty and

Figure 84

he sent his ambassador to Aratta with a new demand: "Let Aratta submit to Uruk!" or else, there will be war!

But the king of Aratta—who might have looked like this statue, found in Harappa, Fig. 84—speaking in a strange language, indicated that he cannot understand what the emissary was saying. Undeterred, Enmerkar sought the help of **Nidaba**, goddess of writing, in inscribing on a clay tablet a written message to Aratta in a language its king would understand, and sent it with another special emissary (the text suggests here that this emissary *flew over* to Aratta: "The herald flapped his wings," and in no time crossed the mountains and reached Aratta).

The inscribed clay tablet—a novelty for the king of Aratta—and

Figure 85

the emissary's gestures, conveyed Uruk's threat. But the king of Aratta put his faith in Inanna: "Inanna, mistress of lands, has not abandoned her House in Aratta, has not handed over Aratta to Uruk!" he said; and the faceoff continued unresolved.

For some time thereafter Inanna shared her presence with both places, commuting between them in her "Boat of Heaven." Sometimes she piloted herself, dressed as a pilot (Fig. 85), sometimes her aircraft was piloted by her personal pilot, Nungal. But prolonged droughts that devastated the grain-based economy of Aratta, and the centrality of Sumer, made Uruk the ultimate winner.

* * *

Several other heroic tales concerning Enmerkar bring into focus the next king of Uruk, ***Lugal.banda***. The King List states laconically, "Divine Lugalbanda, a shepherd, reigned 1,200 years." But considerably more information is provided about him in such texts as *Lugalbanda and Enmerkar, Lugalbanda and Mount Ḫurum,* and *Lugalbanda in the Mountain Darkness*—texts that describe different heroic episodes that could have been segments of an encompassing text—an *Epic of Lugalbanda,* on the pattern of the *Epic of Gilgamesh.*

In one of the tales, Lugalbanda is one of several commanders accompanying Enmerkar on a military campaign against Aratta. As they arrive at Mount Ḫurum on their way, Lugalbanda falls sick. His companions' efforts to help him fail, and they leave him to die, planning to pick up his body on their return. But the gods of Uruk, led by Inanna, hear Lugalbanda's prayers; using "stones that emit light" and "stones that make strong," Inanna restores his vitality and he does not die. He wanders off alone in the wilderness, fighting off howling wild animals, pythons, and scorpions. Finally (presumably, for the tablet is damaged here) he makes his way back to Uruk.

In another tale, he is on a mission from Enmerkar in Uruk to Inanna in Aratta, to seek her help for a water-short Uruk. But in the most interesting version segment, Lugalbanda is depicted as a special emissary of Enmerkar to the king of Aratta. Sent alone on a hush-hush mission with a secret message that he had to memorize, his way is blocked at a vital mountain pass by the "***Anzu Mushen,***" a monster bird whose "teeth are like those of a sharkfish and its claws like a lion's" and who can hunt down and carry a bull. Consistently defined in the text by the determinative ***mushen,*** which means 'Bird', the "Anzu Bird" claims that Enlil placed him there as Gatekeeper, and he challenges Lugalbanda to verify his identity:

> If a god you be,
> the (pass)-word to you I will tell,
> in friendship will I let you enter.
> If a *Lul.lu* you are,
> your fate will I determine—(for)
> no adversary into the Mountainland is allowed.

Bemused, perhaps, by the use of a pre-Diluvial term, Lu.lu, for 'Man',

Lugalbanda answered with his own play of words. Referring to the sacred precinct of Uruk, he said:

> *Mushen,* in the *Lal.u* I was born;
> *Anzu,* in the 'Great Precinct' I was born.

Then "Lugalbanda, he of beloved seed, stretched out his hand" and said:

> Like divine Shara am I,
> ***the beloved son of Inanna.***

The god **Shara** is mentioned in various texts as a son of Inanna, though never with any indication of who his father was. One guess has been that he was conceived during Anu's visit to Earth; the Tale of Zu identifies Shara as "the *firstborn* of Ishtar"—admitting the existence of unnamed others. There is no mention that Inanna's love sessions with Dumuzi produced a child, and it is known that after the death of Dumuzi, Inanna introduced the rite of a "Sacred Marriage" in which a male of her choosing (as often as not the king) would spend a 'betrothal night' with her on the anniversary of Dumuzi's death; but no offspring of record has been listed as a result. That leaves unknown the identity

Figure 86

of Lugalbanda's father, though the inclusion of the term ***lugal*** as part of his name suggests a royal lineage.

It is noteworthy that the meaning of the name ***Lugal.banda*** can best be conveyed by the nickname 'Shorty', for that is what his name literally meant: ***Lugal*** = King, ***banda*** = 'Of lesser/shorter [stature]'. Missing the great size of other demigods, he seems to have been more like his mother in this respect: When a life-size statue of Inanna was discovered at a site called Mari, the archaeologists took a picture of themselves with the statue (Fig. 86); and indeed, Inanna looked the shortest in the group.

Whoever Lugalbanda's father was, the fact that a goddess—Inanna—was his mother earned him the determinative ***Dingir*** before his name, and qualified him to be ***chosen to become the consort of a goddess named*** **Ninsun**. His name, with the ***Dingir*** determinative, concludes the Inanna listings on Tablet IV in the Great God List and is granted the honor of starting Tablet V, followed by ***dNinsun dam bi sal***—'divine Ninsun, female, his spouse'—and by the names of their children and varied court attendants.

Which brings us to the greatest epic tale of demigods and the Search For Immortality—and the existence of physical evidence that might prove it all.

THE CONFUSION OF LANGUAGES

According to the Bible, when people began to resettle the Earth after the Deluge, all of Mankind spoke one language (Genesis 11:1):

> The whole Earth was of one language
> and of one kind of words.

It was so when the people "journeying from the east, found a plain in the land of Shine'ar, and they settled there." But then they started to "build a city and a tower whose head will reach heaven." It was to stop such ambitions on the part of Mankind that Yahweh, having "come down to see the city and the Tower," got concerned and said: "Let us come down and confound there their language, so that they may not understand each other's speech." It was the building of the 'Tower of Babel' that made Yahweh "confuse Mankind's language," and "scatter them from there over the face of the whole Earth."

Then, using a play of words—the similarity of the Hebrew verb *BLL* (= 'confuse, mix up') with the name of the city (*'Babel'* = Babylon)—the Bible explained: "Therefore is the name of it *Babel,* for there did Yahweh *BLL* (= 'confuse') the language of the Earth." The Greek historian Alexander Polyhistor, quoting Berossus and other sources, also tells that before building a large and lofty tower, Mankind "was of one language."

That all of Mankind—stemming from three sons of Noah—would have spoken one language right after the Deluge is a plausible assertion. Indeed, it might explain why the earliest terms and names in Egyptian sound like Hebrew: The word for 'gods' was *Neteru,* "guardians," which matches the Hebrew *NTR* (= 'To guard, to watch'). The name of the chief deity, *Ptah̲,* meaning 'He who develops/creates', is akin to the Hebrew verb *PTH̲* with a similar meaning. The same goes for *Nut* (= 'Sky') from *NTH*—to spread a canopy; *Geb* ('He who heaps up') comes from *GBB* ('to heap up'), etc.

The Bible then states that the Confusion of Languages was *a delib-*

erate divine act. ***Imagine finding corroboration for that in the Enmerkar texts!***

Reporting the inability of Enmerkar's emissaries and the king of Aratta to understand each other, the Sumerian text noted that "once upon a time—

> The whole Earth, all the people in unison.
> To Enlil in one language gave praise.

But then **Enki**, pitting king against king, prince against prince, "put in their mouths a confused tongue, and the language of Mankind was confounded."

According to the Enmerkar epics, Enki did it . . .

XII

Immortality: The Grand Illusion

Once upon a time the whole of Mankind lived in Paradise—satiated from eating the Fruit of Knowledge, but forbidden from reaching for the fruit of the Tree of Life. Then God, mistrusting his own creation, said to unnamed colleagues: The Adam, having eaten of the Tree of Knowing, "has become as one of us; what if he put forth his hand and took also of the Tree of Life, and ate, and lived forever?" And to prevent that, God expelled Adam and Eve from the Garden of Eden.

Man has searched for that God-withheld immortality ever since. But throughout the millennia, it has gone unnoticed that while in respect to the Tree of Knowing *Yahweh Elohim* stated that having eaten of it "The Adam has become as one of us"—no such "as one of us" is asserted in respect to "living forever" from the fruit of the Tree of Life.

Was it because "Immortality," dangled before Mankind as a distinctive attribute of the gods, was no more than a Grand Illusion?

If ever did someone try to find out, it was Gilgamesh, King of Uruk, son of Ninsun and Lugalbanda.

As enchanting and revealing the tales of Enmerkar and Lugalbanda are, without doubt the post-Diluvial *Lugal* and demigod of whom we have the longest and most detailed records is ***Gilgamesh,*** who reigned in Uruk from circa 2750 to circa 2600 B.C. The long *Epic of Gilgamesh* relates his search for immortality—because "two-thirds of him is god, one-third of him is human," and he believed that therefore he should not "peer over the wall" as a mortal.

The genealogical lineage that made him more than a demigod, more than a fifty-fifty god, was impressive. His father, Lugalbanda, both king and high priest in Uruk, was a son of Inanna and was endowed with the "divine" determinative. His mother, **Nin.sun** (= 'Lady Who Irrigates') was a daughter of the great deities **Ninurta** and his spouse **Ba'u**, which explains why Gilgamesh was described as being of the "essence of Ninurta" (Enlil's foremost son). Bau herself was of no mean lineage: She was the youngest daughter of Anu.

That was not the whole pedigree of Gilgamesh. He was born in the presence and under the aegis of the god **Utu** (twin brother of Inanna and a grandson of Enlil)—an aspect that leads scholars to call Utu/Shamash the 'godfather' of Gilgamesh. And he was also "looked upon with favor" on the Enki'ite side, for his full theophoric name, **Gish.bil.ga.mesh**, linked him to d**Gibil**, a son of Enki and the god of metal foundries.

According to a Hittite version of the Epic of Gilgamesh, he was "*lofty, endowed with a super-human size*"—attributes undoubtedly inherited not from the father ('King Shorty') but from the mother's side, for the mother of Ninsun, the goddess Bau, was true to her nickname **Gula**—plain and simple, 'The Big One'.

Bestowed with talents and prowess by several gods, tall, muscular, and shapely (Fig. 87), Gilgamesh was likened to a wild bull; bold and untamed in spirit, he constantly challenged the city's youths to wrestling matches (which he always won). "Unbridled in arrogance," he "left not a maiden alone." Finally the city's elders appealed to the gods to stop Gilgamesh when he started to demand 'first rights' with brides on their wedding night.

Responding, the gods fashioned in the steppe a wild man as a double of Gilgamesh—"Like Gilgamesh in build, though shorter in stature." Called ***Enki.du*** (= 'By Enki created'), his task was to shadow Gilgamesh and force him to change his ways. Discovering that they have on their hands an uncouth Primitive who knows not cooked food and befriends animals, the city's elders put him up outside of town with a harlot, to learn "the ways of Man." She also cleaned and clothed him, and made his hair in curls; when he finally entered town, he was a Gilgamesh duplicate!

Challenged to a wrestling match by an incredulous Gilgamesh, Enkidu wrestled him down and instilled humility in him; and the two became inseparable comrades.

Deprived of his haughtiness and losing his prowess, Gilgamesh began to ponder matters of aging, of life and death. "In my city man dies, oppressed is my heart; man perishes, heavy is my heart," Gilgamesh told his 'godfather' Utu; "Will I also peer over the wall, will I be fated thus?" he asked. The response he got from his mentor was not encouraging:

> Why, Gilgamesh, do you rove about?
> The Life that you seek, you shall not find!
> When the gods created Mankind,
> Death for Mankind they allotted;
> Enduring Life they retained in their own keeping.

Figure 87

Live and enjoy life day by day, Utu/Shamash advised Gilgamesh; but a series of dreams and omens, including a crashed celestial object, convinced Gilgamesh that he could avoid a mortal's end were he to join the gods in their heavenly abode. Enkidu, he learnt, knew the way to the "Landing Place of the Anunnaki" in the Cedar Forest—a great platform with a launch tower, all built of colossal stone blocks, that served as the Earth-terminal for the Igigi and their shuttlecraft (see Fig. 60). It was a place from which he could be taken aloft by the Igigi; and Gilgamesh asked his mother for her advice and help. Told that "only the gods can scale heaven, only the gods live forever under the sun," and warned by Enkidu of the monster Huwawa who guards the place, Gilgamesh answered with words that resonate to this day:

> As for Mankind, numbered are their days;
> Whatever they achieve, is but the wind . . .
> Let me go there before you,
> Let your mouth call out, "Advance! Fear not!"
> And should I fall,
> I shall have made me a name:
> "Gilgamesh," they will say,
> "against fierce Huwawa has fallen."

Realizing that Gilgamesh would not be deterred, Ninsun, his mother, appealed to Utu/Shamash to grant Gilgamesh extra protection. "Wise and versed in all knowledge," Ninsun was also practical. Taking Enkidu aside, she made him swear that he would bodily protect Gilgamesh. To assure his fidelity, she offered him a reward beyond anyone's dreams: *A young goddess as wife.* (The partly damaged lines, at the end of Tablet IV of the Epic, suggest that Ninsun had discussed with **Aya**, spouse of Utu/Shamash, which one of their daughters the bride should be.)

Then Utu/Shamash himself gave Gilgamesh and Enkidu divine sandals that enabled them to reach the Cedar Mountain in a fraction of time, and off were the comrades on their Cedar Forest adventure.

Though no map has been found alongside the ancient text, there is no uncertainty regarding the comrades' destination: In the all of the

Near East—in the whole of Asia—there is only one Cedar Forest: In the mountains of what is now Lebanon; and it was there that the gods' "Landing Place" was located.

Reaching the mountain range, the comrades were awed by the sight of the majestic cedar trees and stopped for the night at the foot of the forest. But during the night Gilgamesh was awakened from his sleep by the shaking of the ground; he managed to glimpse a "sky chamber" lifting off. "The vision that Gilgamesh saw was wholly awesome":

> The heavens shrieked, the earth boomed;
> Though daylight was dawning, darkness came.
> Lightning flashed, a flame shot up.
> The clouds swelled, it rained death!
> Then the glow vanished; the fire went out;
> And all that had fallen was turned to ashes.

The sight and sounds of a rocketship launched was truly awesome; but as far as Gilgamesh was concerned, the night's events confirmed that they had reached the 'Landing Place' of the gods. (A Phoenician coin from a much later time still depicted the site with a rocket poised on its platform, Fig. 88.) At daybreak the comrades began to seek the entrance, careful to avoid "weapon-trees that kill." Enkidu found the

Figure 88

gate; but when he tried to open it, he was thrown back by an unseen force. For twelve days he lay paralyzed.

When he was able to move and speak again, he pleaded with Gilgamesh to give up the attempt to open the gate. But Gilgamesh had good news: While Enkidu was immobilized, he (Gilgamesh) had found a tunnel; it might lead them directly to the command center of the Anunnaki! He persuaded Enkidu that the tunnel was the best way in.

The entrance to the tunnel was blocked by overgrown trees and bushes, soil and rocks. As the comrades began to clear it all, "Huwawa heard the noise and became angry." The guardian of the place, Huwawa was as monstrous as Enkidu had described him: "Mighty, his teeth as the teeth of a dragon, his face the face of a lion, his coming like the onrushing of floodwaters." Most fearsome was his "radiant beam": Emanating from his forehead, "it devoured trees and bushes; none could escape its killing force . . . As a terror to mortals has Enlil appointed him."

With no way to escape, the comrades suddenly heard Utu/Shamash speak to them. Do not run to escape, he told them; instead, let Huwawa come near you, then throw dust in his face! Doing as advised, the comrades managed to immobilize Huwawa. Enkidu struck him, and the monster fell to the ground. Enkidu then "the monster put to death."

With "the way to the secret abode of the Anunnaki opened up," the comrades took time to relax and savor their victory. They stopped to rest by a stream; and Gilgamesh took off his clothes to bathe and refresh himself. Unbeknown to them, the goddess Inanna had been watching it all from her skychamber. Attracted by the king's outstanding physique, the ever-young Inanna made her desire clear as she addressed him:

> Come, Gilgamesh, be thou my lover!
> Grant me the fruit of thy love,
> You be my man,
> I shall be your woman!

Promising him a golden chariot, a magnificent palace, lordship over other kings and princes—Inanna was sure that she enticed Gilgamesh; but answering her, he pointed out that he had nothing to offer her, a goddess, in return; and as to the 'love' she promised—how long did her former lovers last? Listing five of them, Gilgamesh described how Inanna "as a shoe which pinches the foot of its owner" cast them off, one after the other, uncaring once their vigor ran out.

The rebuff enraged Inanna. Complaining to Anu, "Gilgamesh has insulted me!" she said, and asked him to let loose against Gilgamesh ***Gud.anna***—the 'Bull of Anu' or the 'Bull of Heaven'—who roamed in the Cedar Mountain. And though warned by Anu that the beast's release will bring about seven years of famine, Inanna insisted that Anu let it loose.

Forgetting the tunnel and the Landing Place, Gilgamesh and Enkidu ran for their lives.

The magical sandals that Utu gave them enabled them "a distance of one month and fifteen days in three days to traverse." Gilgamesh rushed into the city to mobilize its fighters; Enkidu faced off the monster outside Uruk's walls. Each snort of the Bull of Heaven created a pit into which a hundred fighters fell. But as the Bull of Heaven turned around, Enkidu struck it from behind, and killed it.

Speechless at first, "Inanna to Anu raised a cry," demanding that the slayers of Huwawa and the Bull of Heaven be put on trial. An ancient artist depicted on a cylinder seal (Fig. 89) a gloating Enkidu with the

Figure 89

slain Bull of Heaven, and Inanna addressing Gilgamesh under the sign of the Winged Disc.

Deliberating, the gods' views differed. Having slain both Huwawa and the Bull of Heaven, let Enkidu and Gilgamesh both die, Anu said. Gilgamesh did no slaying, let only Enkidu die, Enlil said. The comrades were attacked by the monsters, so no one should die, Utu said. In the end, Gilgamesh was spared; Enkidu to toil in the Land of Mines was sentenced.

* * *

Still seething from the failed attempt at the Cedar Forest, Gilgamesh did not give up his quest to join the gods in their Celestial Abode. Apart from the Landing Place in the north there was the Spaceport, "where the gods ascend and descend." Focal point of new landing and takeoff facilities built by the Anunnaki to replace the earlier ones destroyed by the Deluge, the Spaceport was located in the sacred Fourth Region of ***Tilmun*** (= 'Place/Land of the Missiles') in the Sinai peninsula. The grand pattern incorporated the pre-Diluvial landing platform in the Lebanon mountains ('A' on map, Fig. 90), required the building of the two great pyramids as guidance beacons in Egypt ('B' on map, Fig. 90), and established a new Mission Control Center ('C' on map, Fig. 90) at the place we call Jerusalem.

Tilmun was a zone forbidden to mortals; but Gilgamesh—"two-thirds of him divine"—figured he might be exempt from the prohibition; after all, it was there that Utnapishtim/Ziusudra, he of Deluge fame, was taken to live! And so was a plan conceived by Gilgamesh for the second attempt to find Immortality. Loathe to see Enkidu gone, Gilgamesh had an idea: The Land of Mines was on the sailing way to Tilmun; let the gods allow him to go there by ship—and he will drop off Enkidu on the way. Once more Ninsun had to make appeals; once again, Utu gave grudging help.

And so it was that the comrades were still alive and together as their ship was passing through the narrow strait leading out of the Persian Gulf (as it is now called). At the narrows, on the shore, they noticed

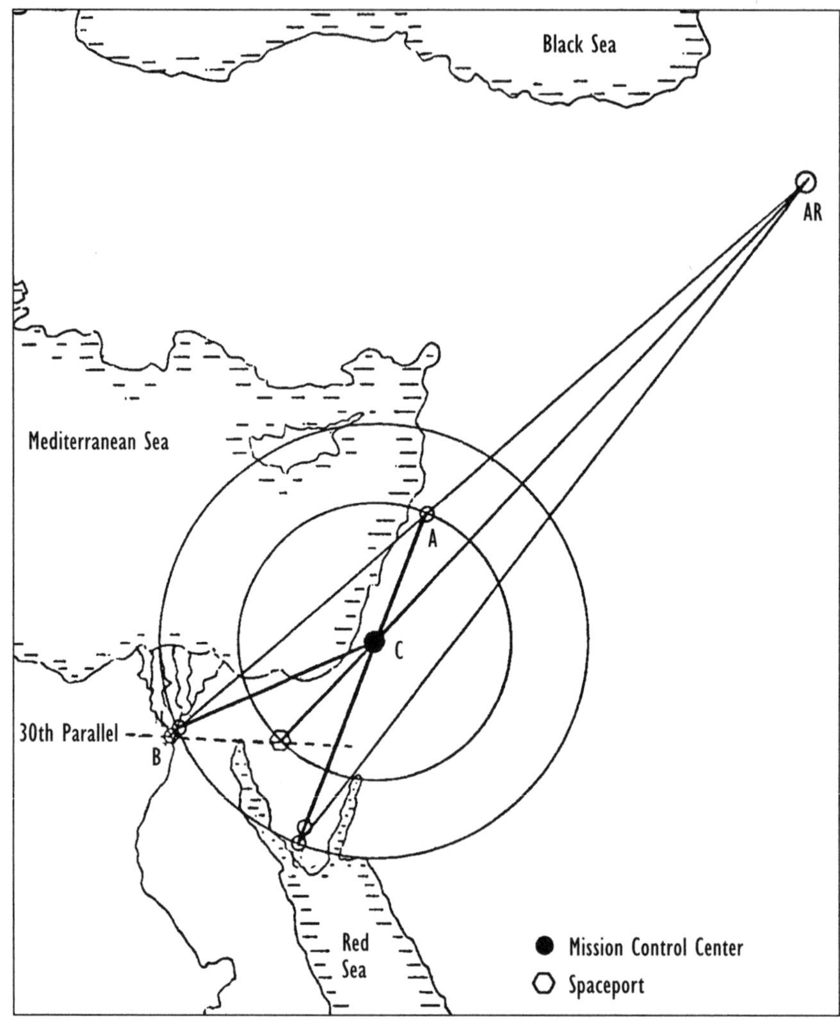

Figure 90

a watchtower. A watchman, armed with a beam like Huwawa's, questioned them. Ill at ease, "Let us turn back!" Enkidu said. "On we go!" Gilgamesh said. A sudden wind, as though driven by the watchman's beam, tore the ship's sail and overturned the boat. In the still darkness of the depths, Gilgamesh saw Enkidu's floating body and dragged it ashore, hoping for a miracle. He sat by his comrade and mourned him day and night, until a worm came out of Enkidu's nostrils.

Alone, lost, and despondent, Gilgamesh at first roamed the wilderness aimlessly; "When I die, shall I not as Enkidu be?" he wondered. Then his self-confidence returned, and "to Utnapishtim, son of Ubar-Tutu, he took the road." Guided by the Sun—he kept going west. At night he prayed to Nannar/Sin, the Moon god, for guidance. One night he reaching a mountain pass; it was the habitat of desert lions, and Gilgamesh wrestled two of them down with his bare hands. He ate their flesh as raw meat, with their skins he clothed himself.

It was an omen that he will overcome all obstacles, Gilgamesh believed; it was also an episode of the epic that artists throughout the ancient world, the Americas included, liked to illustrate as they told the tale (Fig. 91).

Figure 91

* * *

Crossing that mountain range, Gilgamesh could see in the distance below a shimmering body of water. In the adjoining plain he could see a city "closed up about"—surrounded by a wall. It was a city "whose temple to Nannar/Sin was dedicated"—the city known from the Bible as *Yeri<u>h</u>o* (= 'Moon city'), *Jericho* in English. He had reached, the text explains later, the Salt Sea (the 'Dead Sea' in current English, *Yam Hamela<u>h</u>,* 'The Sea of Salt', in the Bible).

Outside the city, "close by the low-lying sea," there was an inn, and Gilgamesh set his steps toward it. The Ale-Woman, *Siduri,* saw him coming and prepared a bowl of porridge; but as he came closer she was frightened, for he wore skins and his belly was shrunk. It took some time for her to believe his story that he was a famed king, looking for his long-living ancestor. "Now, ale-woman," Gilgamesh said, "which is the way to Utnapishtim?" It is a place beyond the Salt Sea, Siduri said, adding that

> Never, O Gilgamesh, has there been a crossing!
> From the Beginning of Days
> none who came could cross the sea—
> Only valiant Shamash crosses the sea!
> Toilsome is the crossing,
> desolate is the way thereto,
> Barren are the Waters of Death
> which it encloses.
> How then, Gilgamesh, will you cross the sea?

With no answer, Gilgamesh remained silent. Then Siduri spoke up again. There is, after all, a way to cross the Sea of the Waters of Death: Utnapishtim has a boatman who comes across from time to time for supplies; *Urshanabi* is his name; go, let him see your face—he might take you across on a raft made of logs.

When the boatman, Urshanabi, arrived, he too (like the ale-woman earlier) found it difficult to believe Gilgamesh that he was once king of Uruk, and Gilgamesh had to tell him the whole tale of his search for immortality, the adventures at the Landing Place, the death of Enkidu, and his wanderings in the wilderness, ending with his encounter with the ale-woman, omitting nothing. "I ranged and wandered over all the lands, I traversed difficult mountains, I crossed all the seas," he said, so that "now I might come and behold Utnapishtim, whom they call The Faraway."

Finally persuaded, the boatman took him across and advised him to proceed in the direction of "the Great Sea, which is in the Faraway." But he had to make a turn when he reached two stone markers, go to a town (called *Ulluyah* in a Hittite rendering), and obtain there permission to continue to Mount *Mashu*.

Following the directions, but cutting short a stay in Ulluyah, Gilgamesh proceeded to Mount Mashu, only to discover that it was more than a mere mountain:

> Rocket-men guard its gateway;
> their terror is awesome, their glance is death.
> Their glaring beam sweeps the mountains;
> They watch over Shamash
> as he ascends and descends.

"When Gilgamesh beheld them, with fear and terror was darkened his face"—and no wonder, judging by the way ancient illustrators depicted them (Fig. 92). The guards were just as surprised; as a Rocket-man's beam swept its glare over Gilgamesh, with no apparent effect, he called to his fellow guard: 'He who approaches us, his body is the flesh of the gods! Two-thirds of him is god, one-third is human!'

"Why have you come here?" they challenged Gilgamesh, "the purpose of thy coming we need to learn." Regaining his composure, he approached them. "On account of Utnapishtim, my forefather, who joined the Assembly of the gods have I come, about Death and Life I wish to ask him," Gilgamesh answered.

"Never was there a mortal who could achieve that!" the Rocket-man said, telling him of Mount Mashu and the underground passageway to it. "The mountain's trail no one has traveled; for twelve leagues extends

Figure 92

its interior; dense is the darkness, light there is none!" But Gilgamesh was not dissuaded, and the Rocket-man "the gate of the mountain opened for him."

For twelve double-hours Gilgamesh advanced in the tunnel in darkness, feeling a fresh air breeze only at the ninth hour; a faint light appeared in the eleventh double-hour. Then he walked out into brightness and an incredible sight: an "enclosure of the gods" in which there grew a 'garden' made entirely of precious stones—

> As its fruit it carries carnelians,
> its vines too beautiful to behold.
> The foliage is of lapis lazuli;
> the grapes, too luscious to look at,
> of [. . .] stones are made.
> Its [. . .] of white stones [. . .],
> In its waters, pure reeds [. . .] of *Sasu*-stones
> Like a Tree of Life and a Tree of [. . .]
> that of *An.gug* stones are made.

As the description goes on, it becomes clear that Gilgamesh found himself in an artificial Garden of Eden, made entirely of precious stones. Gilgamesh was marveling at the sight when he suddenly saw the man he went searching for, the "One of the Faraway." Coming face to face with an ancestor from millennia past, this is what Gilgamesh had to say:

> As I look upon thee, Utnapishtim,
> Thou art not different at all,
> even as (though) I am thou . . .

Then, telling Utnapishtim of his search for Life and the death of Enkidu, he said to him, to Utnapishtim:

> Tell me,
> How did you join the congregation of the gods
> in thy quest for Life?

Well, Utnapishtim said, it was not that simple. A secret of the gods let me tell you, he said:

> Once, the Anunnaki, the great gods, convened;
> *Mammetum,* maker of Fate,
> with them the fates determined . . .
> Shuruppak, a city which thou knowest,
> a city which on the Euphrates is situated,
> that city was ancient, as were the gods within it.
> When their heart led the great gods to the Deluge,
> the Lord of Pure Foresight, Ea, was with them.
> Their words he repeated (to me) through the reed wall:
> "Man of Shuruppak, son of Ubar-Tutu,
> Tear down house, build a ship!
> Give up possessions, seek thou Life!
> Aboard the ship take the seed of all that lives."

Describing the ship and its measurements, Utnapishtim went on to tell Gilgamesh that the townspeople of Shuruppak helped build the ship for they were told that they would thereby get rid of Utnapishtim, whose god was quarreling with Enlil. Telling the whole story of the

Deluge, Utnapishtim related how Enlil discovered Ea/Enki's duplicity, and how Enlil, changing his mind, blessed Utnapishtim and his wife to live henceforth "the life of the gods":

> Standing between us,
> he touched our foreheads to bless us:
> "Hitherto, Utnapishtim has been human;
> henceforth, Utnapishtim and his wife
> like gods shall be unto us.
> Faraway shall the man Utnapishtim reside,
> at the mouth of the water-streams."

"But now," Utnapishtim went on to say to Gilgamesh, "who will for thy sake call the gods to Assembly, that the Life that thou seekest thou mayest find?"

Hearing that, and realizing that his search has been in vain, for only the gods, in assembly, can grant Eternal Life—Gilgamesh fainted, lost consciousness, and collapsed.

* * *

For six days and seven nights Utnapishtim and his wife kept vigil as Gilgamesh slept uninterrupted. When he finally awoke, with the help of Urshanabi they bathed Gilgamesh and dressed him with clean garments as befits a king returning to his city. It was at the very last moment that Utnapishtim, pitying Gilgamesh as he was leaving empty handed, suddenly said to him: "What shall I give thee, as you return to thy land?" He had a going-away gift for him, a "secret of the gods":

> To you, O Gilgamesh,
> a hidden thing I will disclose—
> A secret of the gods I will tell thee:
> A plant there is, like the buckthorn's is its root.
> Its thorns are like a brier-vine's,
> thy hands they will prick.
> (But) if thy hands obtain the plant,
> New Life thou shall find!

The rejuvenating plant, Utnapishtim said, grows at the bottom of a water hole (or well)—and showed Gilgamesh where. "No sooner did Gilgamesh hear this, than he opened the water pipe. He tied heavy stones to his feet; they pulled him down into the deep, and he saw the plant. He grabbed the plant, though it pricked his hands. He cut the heavy stones from his feet; the well cast him up by its shore."

Holding on to the rejuvenating plant—a scene possibly depicted on an Assyrian monument, Fig. 93—the overjoyed Gilgamesh spilled out to Urshanabi, the boatman, his future plans:

> Urshanabi, this plant is a plant unlike any,
> Whereby a man may regain his life's breath!

Figure 93

I will take it to ramparted Uruk,
I will cause [. . .] to eat the plant [. . .],
'Man Becomes Young In Old Age'
its name shall be.
I myself shall eat (of it),
and to my youthful state I shall return!

Certain that he had finally attained his life's dream, Gilgamesh started on the way back to Uruk, accompanied by Urshanabi. After twenty leagues Gilgamesh and Urshanabi "stopped for a morsel." After another thirty leagues, "they saw a well and stopped for the night." Filled with visions of rejuvenation, Gilgamesh put down the bag with the unique plant to take a refreshing swim; and while he was not watching,

A serpent sniffed the fragrance of the plant;
It came up from the water and carried off the plant.
And Gilgamesh sat down, and wept,
His tears running down his face.

Gilgamesh the demigod cried, for once again Fate had snatched for him defeat out of success. Mankind, one believes, has cried ever since—for this was the greatest irony of all: It was the Serpent who encouraged Mankind to eat of the Forbidden Fruit without fear of dying—and it was the Serpent who robbed Man of the Fruit of Not Dying . . .

Was it again a metaphor for Enki?

* * *

Gilgamesh, the Sumerian King List says, reigned for 126 years and was followed on the throne by his son ***Ur.lugal.*** His death, as his whole tragic tale, leaves unanswered the question that is its central theme: Can Man—even if he be partly god—avoid mortality? And if his life was an unanswered puzzle, his death was even more so when it comes to his burial.

From Gilgamesh in the 3rd millennium B.C. to Alexander in the 4th century B.C. to Ponce de Leon (searching for the Fountain of

Youth) in the 16th century A.D., Man has searched for a way to avoid, or at least postpone, dying. But is that universal and ongoing search the opposite of what Man's creators had planned? Do the cuneiform texts and the Bible imply that the gods *deliberately* held back immortality from Man?

In the Epic of Gilgamesh the answer is a statement that is given as fact, amounting to a Yes:

> When the gods created Mankind,
> Death for Man they allotted—
> 'Everlasting life'
> they retained in their own keeping.

Gilgamesh heard it from his godfather Utu/Shamash, when his interest in Life and Death matters began, and once more from Utnapishtim (after Gilgamesh told him his journies' purpose). The answer is: It's a useless effort—Man cannot escape his mortality, and the whole long tale of Gilgamesh seems to confirm that.

But let us re-read the tale, and the irony in that apparent answer emerges: The way to attain the longevity of the gods, his mother told Gilgamesh, was to *join them on their planet*. That explains why the same Utu/Shamash who at first said 'Forget it', then provided help to Gilgamesh on his two attempts to go where the rocketships ascend and descend. Failing that, a "secret of the gods" was revealed to Gilgamesh: The existence of a rejuvenating Plant of Life right here on Earth. And that raises the question about the gods themselves: Did *their* "everlasting life" also depend on such a nutrient—were they not the wonted 'Immortals'?

Interesting light is cast on the subject from ancient Egypt, where the Pharaohs held the belief that everlasting life awaits them in an Afterlife if they could join the gods on the "Planet of Millions of Years." To achieve that, elaborate preparations were made ahead of time to facilitate the Pharaoh's journey after his death. Starting with the exiting by the Pharaoh's *Ka* (a kind of an Afterlife Alter Ego) from his tomb via a simulated door, the king journeyed to the *Duat* in the Sinai peninsula, there to be taken aloft for a space trip. (The existence of such a facility in the Sinai was attested by a tomb depiction showing a multistage rocketship [akin to the Sumerian ***Din.gir*** symbol!] in an underground silo, Fig. 94.) Detailed text and drawings in the *Book of the Dead* then described the subterranean facilities, the rocketships' pilots, and the breath-taking liftoff.

But the purpose of the space journey was not to merely reside on the gods' planet. "Take ye this king with you, that he may eat of that which you eat, that he may drink of that which ye drink, that he may live on that whereupon you live," an ancient Egyptian incantation appealed to the gods. In the pyramid of King Pepi, an appeal was made to the gods whose abode was on the "Planet of millions of years" to "Give unto Pepi the Plant of Life on which they themselves are sustained." A colorful drawing on Pyramid walls showed the king (here accompanied by his wife) arriving in the Afterlife at the Celestial

Figure 94

Paradise, sipping the Water of Life out of which there grows the Tree of the Fruit of Life (Fig. 95).

The renderings from the Egyptian side regarding the gods' Water of Life and Food/Fruit of Life match the Mesopotamian depictions of Winged Gods ("Eaglemen"), flanking the Tree of Life as they hold in one hand the Fruit of Life and in the other a pail of the Water of Life (see Fig. 72). The notions underlying these depictions are no different from the Hindu tales of the *Soma*—a plant that the gods had brought to Earth from the heavens—whose leaves' juice conferred inspiration, vitality, and immortality.

While all that appears to be in accord with the biblical take on the subject, which is best known from the tale of the two special trees in the Garden of Eden—the Tree of Knowing and the Tree of Life, whose fruit could make Adam "live for ever"—the biblical tale also relates *a divine effort to prevent Man from partaking in that fruit.* Man was expelled from Eden "lest he try," and God was so determined to prevent the Earthlings from regaining access to the Tree of Life, that He "placed at the east of the Garden of Eden the *Cherubim* and the flaming sword which revolveth, to guard the way to the Tree of Life."

The tale's essential element—of Man's creator trying to prevent him from having divine nourishments—is found in the Sumerian tale

Figure 95

of Adapa. There we find the Creator of Man himself, Enki, treating the "perfect model of Man," his own Earthling son Adapa, thus:

> Wide understanding he perfected for him;
> Wisdom he had given him;
> To him he had given Knowledge—
> ***Lasting life*** he had not given him.

Then Enki's own handiwork is put to the test: Adapa, his son by an Earthling woman—a marvel invited by Anu to Nibiru—is offered there the "Food of Life" and the "Water of Life," but is told by Enki to avoid both, for they will cause Adapa's death. That, it turns out, is not true—just as God's warning to Adam and Eve that eating the fruit of the Tree of Knowing would cause death was not true. What worries God (in the Garden of Eden tale) is not the couple's risk of death but the opposite—"What if the Adam put forth his hand and took also of the Tree of Life, and ate of it, and *lived forever*?"

(The Hebrew words in the Bible are *Ve'akhal ve Chai* ***Le'Olam***— "and he ate and lived ***to OLAM***." The term *Olam,* usually translated "forever, everlasting," etc., can also refer to a physical place, in which case *Olam* is translated 'World'. It can also stem, I have suggested, from the verb that means 'to vanish, to be unseen', so that ***Olam*** could be ***the Hebrew name for Nibiru,*** and in this context, the longevity place. See the sidebar "Words and Their Meaning," page 197.)

God, thus, was worried that were The Adam to eat of the Tree of Life, he would gain the life span "*of* Olam', the life cycle of Nibiru. In the Sumerian text Enki tricks Adapa not to have the divine nourishments simply because when Man was created, Everlasting Life was deliberately held back from him. While the existence of a Food of Life is affirmed, it is ***not Immortality***—it is "Lasting Life," ***Longevity***—that has been deliberately held back from Man. The two may have the same short term result, but they are not the same thing.

Now, what was that "Life of Olam," life on Nibiru—an endless immortality or simply a great longevity that on Nibiru is counted in *Shar* units—3,600 times longer than Earth's life cycles? The notion of gods (or even demigods) as Immortals has come to us from Greece; the discovery in late 1920s of Canaanite 'myths' at their capital Ugarit (on Syria's Mediterranean coast) showed how the Greeks got the idea: From the Canaanites, via the island of Crete.

But in Mesopotamia, the Anunnaki gods never claimed absolute Not Dying—Immortality—for themselves. The very listing of earlier generations on Nibiru amounts to saying: Those were the forebears who have died. The tale of **Dumuzi** was a public telling of his death, a death recorded and mourned (even in Jerusalem at the time of the Prophet Ezekiel) on the anniversary month of *Tammuz*. **Alalu** was sentenced to die in exile; **An.Zu** was executed for his crime; ***Osiris*** was killed and dismembered by Seth; the god ***Horus*** died from a scorpion's sting (but was revived by Thoth). Inanna herself was seized and put to death when she entered the Lower World without permission (but was revived through Enki's efforts).

There was no immortality, nor even a claim of immortality by the Anunnaki. ***There was an illusion of immortality,*** caused by ***extreme Longevity.***

That longevity was apparently associated with living on Nibiru and not just being sustained by some of Nibiru's unique nourishments, for otherwise what purpose was there for Ninsun to encourage Gilgamesh to go there, to attain the "Life of a god."

An interesting question for modern science to ponder is this: Was the life cycle longevity (on Nibiru, or anywhere else in the universe) an aqcuired trait, or an evolutionary genetic adjustment? The statement associated with Adapa suggests a genetic decision by Enki—that a '***Longevity Gene***' (or genes), ***known to Enki,*** was deliberately excluded from the human genome when the 'mixing' of genes took place. Could we ever find it?

That a key with which those genetic secrets could be unlocked might be available, is where our trail of gods and demigods is leading.

SPELLING OUT 'LIFE'

The King James translators of the Hebrew Bible, and virtually all who followed them, have done their best to instill a sense of Divine Spirit, a majestic awe of a Creator of All, in the creation stages described in Genesis. The "winds"—satellites—of Nibiru/Marduk, ***Ru'a<u>h</u>*** in Hebrew, become the *Spirit* (of God) hovering over dark chaos; the *Elohim,* fashioning The Adam "in our image and after our likeness," breathe the "breath of life" (***Neshamah***) into his nostrils and give him a "*soul.*"

On our way we have stopped here and there in this book to (a) point out misconceptions stemming from mistranslations, and (b) to highlight instances where the Hebrew is a literal rendering of a Sumerian term, identifying source word by word and making understanding the verse clearer.

It was the noted Sumerologist Samuel Noah Kramer who had pointed out that in the tale of fashioning Eve out of Adam's *rib*—***<u>Ts</u>ela*** in Hebrew—the Hebrew redactor must have taken the Sumerian word ***Ti*** to mean "rib"—correctly, except that a similarly pronounced word ***Ti*** in Sumerian meant "Life," as in **Nin.ti** (= 'Lady Life'): What was done was to take that which is "Life"—DNA—from the Adam and manipulate it to obtain a female genetic chromosome.

These instances come to mind as one reads the actual Sumerian wording used by Ziusudra to tell Gilgamesh how Enlil granted him the "Life of a god":

Ti Dingir.dim Mu.un.na Ab.e.de
Zi Da.ri Dingir.dim Mu.un.na Ab.e.de

Two Sumerian terms, ***Ti*** and ***Zi,*** both usually translated 'Life', were used here; so what was the difference? As best as one can determine, ***Ti*** was used to indicated the *physical* godlike aspects; ***Zi*** expressed Life's *functioning,* how the living is carried out. To make his meaning extra clear, the Sumerian author added the term ***Da.ri*** (= 'Duration') to ***Zi;*** what Ziusudra was granted was both the physical aspects of godly life, as well as the *durability* aspects of it.

The two lines are usually translated "*Life* like that of a god he gives to him, an *eternal soul* like that of a god he creates for him."

A masterful translation, to be sure, but not the exact meaning of the Sumerian writer's masterful play of words, using ***Ti*** once and ***Zi*** (as in Ziusudra) in the next line. ***Not a 'soul', but* durability, *was added to Ziusudra's Life.***

XIII
Dawn of the Goddess

"Come Gilgamesh, be thou my lover!"

There are hardly any other few words, here spoken by Inanna, that epitomize the unintended consequences of the post-Diluvial relationship between gods and Earthlings.

In truth, after it was realized that the Anunnaki could have all the gold they needed just by picking it up in the Andes, there was no reason for them to stay on in the Old Lands. Enlil, according to Ziusudra, changed his mind about the imperative of wiping Mankind off the face of the Earth after he smelled the aroma of roasting meat—the thanksgiving sacrifice of a lamb offered by Ziusudra; but in fact the change of heart among the Anunnaki leadership began as soon as the scope of the calamity became clear.

While down below the avalanche of waters swept everything away, the gods were orbiting the Earth in their aircraft and shuttlecraft. Crammed inside, "the gods cowered like dogs, crouched against the walls . . . the Anunnaki were sitting in thirst, in hunger . . . Ishtar cried out like a woman in travail; the Anunnaki gods wept with her: 'Alas, the olden days are turned to clay'." Most touched was Ninmah:

> The great goddess saw and wept . . .
> Her lips were covered with feverishness.
> "My creatures have become like flies—

they fill the rivers like dragonflies,
their fatherhood taken away by the rolling sea."

When the tidal wave retreated and the twin peaks of Mount Ararat emerged from the endless sea, and the Anunnaki began to bring their craft down, Enlil was shocked to discover the survival of 'Noah'. Long verses detail the accusations hurled at Enki once his duplicity came to light, and his justification of what he had done. But equally long verses record the vehement reprimand that Ninmah̲ directed at Enlil for his "Let's wipe them off" policy. We have created them, now we are responsible for them! she, in essence, said; and that, plus the realities of the situation, convinced Enlil to change his mind.

Ninmah̲—a female of 'Shakespearean' dimensions, were he to live in her time—had played major roles in the affairs of gods and men before the Deluge; and she did so, though in different ways, thereafter too. A daughter of Anu, she was caught in a love triangle with her two half-brothers, having an out-of-wedlock child (Ninurta) with Enlil after she was prevented from marrying Enki whom she loved. Considered important enough to be granted one of the first five pre-Diluvial cities (Shuruppak), she came to Earth to serve as Chief Medical Officer of the Anunnaki (see Fig. 65), but ended up creating ***Ameluti***—workmen—for them (earning her the epithets **Ninti, Mammi, Nintur,** and many more). Now she saw her creatures turned to clay, and she raised her voice against Enlil.

Thereafter, she was the arbiter between the rival half-brothers and their clans. Respected by both sides, she negotiated the peace terms that ended the Pyramid Wars and was granted the sacred Fourth Region (the Sinai Peninsula) with its Spaceport as neutral territory. A long text describes how her son Ninurta created a comfortable abode for her amidst the mountains of the Sinai peninsula, resulting in her Sumerian name **Ninh̲arsag** (= 'Lady/Mistress of the Mountain peak') and the Egyptian epithet *Ntr Mafqat* (= 'Goddess/Mistress of Turquoise', which was mined in the Sinai). She was worshipped in Egypt as the goddess Hathor (literally *H̲at-H̲or,* 'Abode of Horus'), and in her old age was nicknamed 'The Cow' both in Sumer and in Egypt, for her asserted role in breastfeeding demigods. But at all times, whenever the title 'Great Goddess' was used, it was always reserved for her.

Never married—the original "Maiden" of the zodiacal constellation we call Virgo—she had, in addition to her son with Enlil, several daughters by Enki born on Earth as a result of lovemaking on the banks of the Nile. The tale, that has been misnomered *A Paradise Myth,* ends with Ninharsag and Enki engaged in matchmaking, pairing off young goddesses with Enki'ite males; prominent among them were spouses chosen for **Ningishzidda** (Enki's science-knowing son) and for **Nabu** (Marduk's son)—powerful matchmaking feats, to be sure; but as we shall see, not the last of Ninharsag's power-links and string-pullings through births and marriages, in which she was joined by her younger sister, the goddess **Ba'u**, and by Bau's daughter **Ninsun**.

Bau, who had also come from Nibiru, was one of the Anunnaki female 'great gods'. She was the spouse of Ninurta, which made her daughter-in-law of Ninharsag. But Bau herself was the youngest daughter of Anu, which made her a sister of Ninharsag . . . Both ways, these relationships served as a special bond between the two goddesses, especially so since Bau too gained a reputation as a medical doctor, credited in several tales of bringing the dead back to life.

When she and Ninurta settled in a new sacred precinct that a king of Lagash, Gudea, had built for them, the place became a kind of field hospital for the people (rather than for gods)—a unique aspect of the love for humankind that Bau picked up from Ninharsag. Lovingly nicknamed ***Gula*** (= 'The Big One'), she was invoked in prayers as "Gula, the great physician"—and in curses was asked to "put illness and unhealing sores" on an adversary. The nickname, in any event, correctly invoked her hefty size (see Fig. 80).

If Ninmah/Ninharsag was the first "always a bridesmaid but never a bride," Ninsun, her granddaughter (via Ninurta) cum niece (via Bau), was "always a bride" (in a manner of speaking), for a line of renowned kings claimed to have been her sons—among them the great Gilgamesh. Starting with his father and continuing into the Third Dynasty of Ur and beyond, she outlived one mortal spouse after another. Her family album (were she to have one) bulged with children and grandchildren—starting with her own eleven children with the deified demigod Lugalbanda.

The three—Ninẖarsag, Bau, Ninsun—formed a trio of goddesses who had a hand in steering Sumer's royals in life as well as in death (including *the most challenging female mystery*). A fourth principal female activist—Inanna/Ishtar—had, as we shall see, her own agenda.

* * *

Having reconciled to sharing the Earth with Mankind, the Anunnaki set out to make Earth habitable again after the Deluge. In the Nile Valley, Enki—***Ptah*** to the Egyptians—built dams with sluices (see Fig. 12) to drain off floodwaters and, in the words on a papyrus, "to lift the land from under the waters." In the Euphrates-Tigris plain, Ninurta created habitable areas by damming mountain passes and draining the water overflow. At a "Chamber of Creation"—in all probability situated on the great stone platform that the Igigi had used as a 'Landing Place'—Enki and Enlil supervised feats of genetic 'domestication' of plants and animals. The zeal with which all that was done suggests that the Anunnaki leaders were captivated by their own vision of becoming Interplanetary Benefactors. Right or wrong, they did create the Earthlings, who served them well as toilers in mines and fields; so Anu's state visit to Earth circa 4000 B.C. put in motion a decision that it was only right to give Mankind 'Kingship'—Civilization—by rebuilding pre-Diluvial cities (exactly where they had been) and establishing several new ones.

Much has been written, based on archaeological discoveries, of how cities became 'cult centers' for this or that particular deity, with an '***E***' ('Abode' = Temple) in a 'sacred precinct' where priests provided the resident deities with the leisurely life of privileged overlords. But not enough has been written of the role of 'at large' deities who were the mainstay of civilized advancement: A deity, **Nidaba**, who was in charge of Writing, overseeing regular as well as specialized scribal schools; or **Nin.kashi**, who was in charge of the beer-brewing that was one of Sumer's 'firsts' as well part of its social life; or **Nin.a**, who supervised the land's water resources.

Those deities were *goddesses;* so was **Nisaba**, also known as ***Nin.mul.mula*** (= 'Lady of many planets' or 'Lady of the Solar System'), an astronomer whose tasks included providing celestial orientation for new temples—not only in Sumer but also in Egypt (where she was revered as *Sesheta*). Another female deity, the goddess **Nanshe**, was mistress of the calendar who determined New Year's Day. Added to the 'traditional' medical services provided by the group of *Suds* (= 'One who gives succor') who arrived with Ninmah̲, the specialties overseen by goddesses embraced every aspect of civilized life.

The increased and more assertive role of goddesses in the affairs and hierarchy of the Anunnaki was expressed graphically at a sacred Hittite site called Yazilikaya in central Turkey, where the pantheon of twelve leading deities, carved on rock faces, is depicted as two equal groups of male gods and female goddesses marching toward each other with their retinues (Fig. 96, partial view).

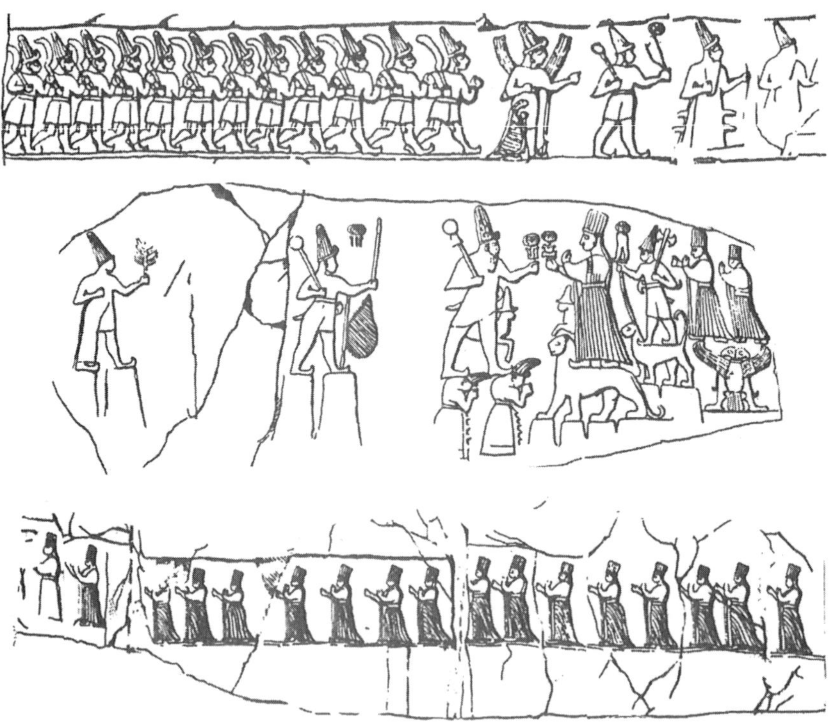

Figure 96

In the relations between Anunnaki and Earthlings, the increasing 'feminization' was enhanced by the actual power and authority wielded by the second and third generations of Anunnaki on Earth. In the Olden Days, the nurse **Sud** was re-titled **Nin.lil** when she became Enlil's spouse, but her title (= 'Lady of the Command') did not make her an Anunnaki commanding leader. Ea's spouse, Damkina, was re-titled **Nin.ki** (= 'Lady [of] Earth') when he was renamed **En.ki**, but she never was Mistress of the Earth. Even **Nin.gal**, spouse of Enlil's Earthborn son **Nannar/Sin**, who in official 'portraits' (Fig. 97) shared equal status with him, had no known authority/powers of her own.

Things were different when it came to goddesses born on Earth, as shown by Nannar/Sin's and Ningal's daughters **Ereshkigal** and **Inanna**. When Inanna was granted Uruk, she turned it into a powerful capital of Sumer; when Marduk caused the death of her bridegroom Dumuzi, she launched and led an intercontinental war; when she was made divine head of Aratta, she insisted that it be granted the full status of the Third Region. She could and did select kings (and ordered them around).

Figure 97

When Ereshkigal (= 'Scented Mistress of the Great Land') was less than enthusiastic about marrying Enki's son Nergal, who was bald and had limped from birth, she was promised to become Mistress of his African domain; called the 'Lower World', it was the continent's southern tip. Ereshkigal made it into the site for crucial scientific observations involving the Deluge and (in subsequent times) of determining zodiacal ages. Text after text describe the ruthless determination with which Ereshkigal wielded the resulting powers.

And one key area in which all these changes came to the fore was the issue of demigods.

With the institution of Kingship came the function and persona of a 'King'—a ***Lu.gal***—"Big Man." Residing in his own ***E.gal,*** the Palace, he ran the administration, promulgated laws, dispensed justice, built roads and canals, maintained relations with other centers, and enabled society to function—all on behalf of the gods. It was, by and large, a formula for growth, achievements in technology and arts, prosperity. As begun in Sumer some 6,000 years ago, it laid the foundations for all that we call Civilization to this day.

It was only natural that someone would come up with the idea that the best *Lu.gal* would be akin to the demigods who were around before the Deluge "and thereafter too." Endowed (in fact or by presumption) with more intelligence, physical strength and size, and longevity than the average Earthling, 'demigods' were the best choice to serve as the link between gods and mortals—to be the kings, especially so when the king also served as the high priest allowed to approach the deity.

But where would those post-Diluvial demigods be coming from? The answer, extracted from varied texts, is this: They were made to order . . .

* * *

With a few exceptions, the Sumerian King List provides no direct information on the demigod status of the kings who made up the First Dynasty of Kish, the one that started post-Diluvial Kingship under the aegis of Ninurta.

Like the King List, we have dwelt on ***Etana*** and his legendary space trips, concluding that his reign length (1,560 years) and eligibility for space visits to Nibiru indicate his demigod status, which is further corroborated by a notation in another text that Etana was of the same "Pure Seed" as Adapa. We have also pointed out that some of the names of subsequent kings of Kish, such as ***En.me.nunna*** (660 years) and ***En.me.bara.ge.si*** (900 years), suggested the presence of demigods in between their non-divine successors. In Tablet I of the Great God List, following the Enlil group and the Ninurta listings, there are fourteen names that start with ***ḏ.Lugal***—*divine* Lugal.gishda, *divine* Lugal.zaru, etc. Unknown otherwise, they represent demigods—entitled to the *din-gir* determinative!—who either did not reign in Kish or were known by other epithet-names.

Where data is provided, we find a major change in 'demigodness'. In pre-Diluvial times, and for a while thereafter, 'demigodness' stemmed from the "Pure Seed" of a male parent: So and so was the son of ḏUtu, etc. It was thus quite a change when a king named ***Mes.Alim*** (also written 'Mesilim')—a name whose significance we shall soon explore—ascended the throne of Kish. One of the discovered artifacts (a silver vase) bears this telltale inscription:

Mes-Alim
king of Kish
beloved son
of ḏNinẖarsag

Since there is no way the king—proved correct in all his other inscriptions—would have dared present the vase to the goddess if it were not true, a birth involving Ninẖarsag as the mother has to be considered in spite of her advanced age; this could include *artificial insemination,* which was in fact claimed in another instance in which Ninẖarsag was involved.

That such pre-assuring from birth the 'demigod qualifications' of a future king was practiced by the Anunnaki is documented by a long and clearly written inscription regarding a king named ***Eannatum*** in the city of Lagash (whose patron god was Ninurta, here renamed **Nin.Girsu** after the city's sacred precinct). Reigning circa 2450 B.C. (by one chronology) Eannatum attained fame as a fierce warrior whose feats were recorded both in texts and on monuments, leaving no doubt about his historicity. On a stela now on exibit in the Louvre (Fig. 98) he claimed ***divine ancestry through artificial insemination*** and a birth involving several deities. Here is what the inscription said:

> Divine Ningirsu, warrior of Enlil,
> implanted the semen of Enlil for Eannatum
> in the womb of [?].
> [?] rejoiced over Eannatum.

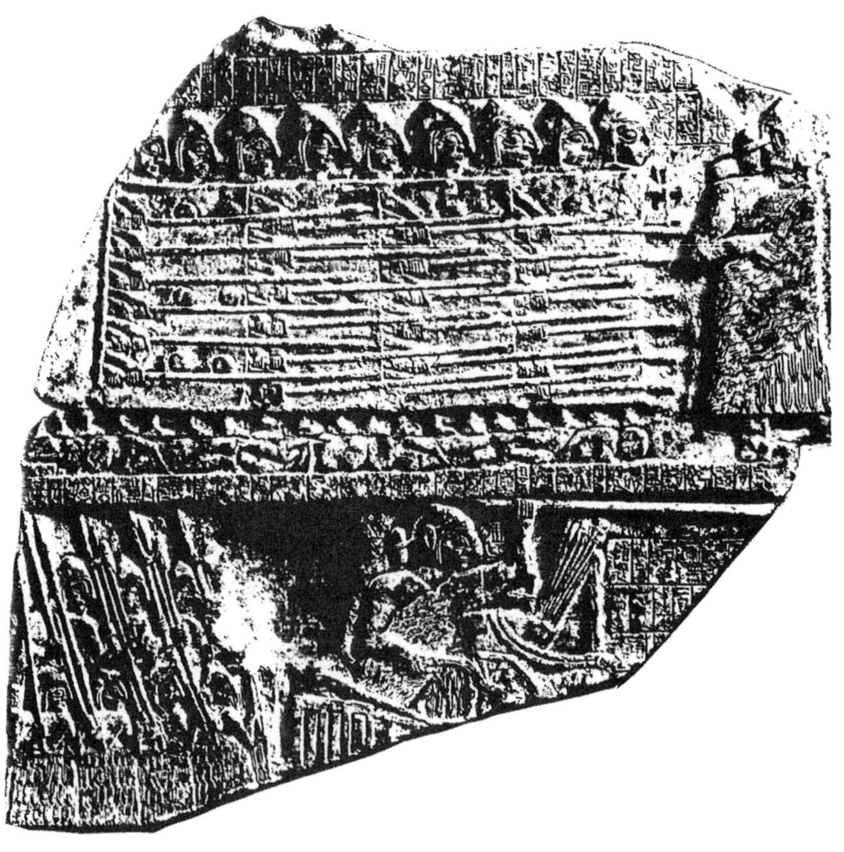

Figure 98

> Inanna accompanied him, named him
> 'Worthy of the temple of Inanna in Ibgal',
> and sat him on the holy lap of Ninharsag.
> Ninharsag offered him her special breast.
> Ningirsu rejoiced over Eannatum,
> semen implanted in the womb by Ningirsu.

As if to answer a future question, the inscription went on to describe the giantlike size of Eannatum:

> Ningirsu laid his extent upon him:
> For a span of five forearms
> he set his forearm on him—
> A span of five forearms for him he measured.
> Ningirsu, with great joy,
> gave him the kingship of Lagash.

(The term 'forearm', usually translated 'cubit', represents the distance from the elbow to the end of the midfinger, on the average about 20 inches. Eannatum's 'span' of five forearms means he was about 100 inches, or over 8 feet, tall.)

An instance of artificial insemination is also recorded in Egyptian tales of the gods, when the god ***Thoth*** (**Ningishzidda** in Sumer) extracted semen of the dead (and dismembered) Osiris and impregnated with it Isis, the wife of Osiris (who then gave birth to the god Horus); a depiction of the tale (Fig. 99) shows Thoth combining two separate strands of DNA to attain the feat. In the case of Eannatum we have a clearly described similar instance—in Sumer—in which the Foremost Son of Enlil was involved. The opening statement regarding "semen of Enlil" is understood to mean Ninurta's own semen, carrying as it did the Seed of Enlil.

Eannatum was followed on the throne in Lagash by king ***Entemena;*** and though stated in inscriptions to have been "son of Eannatum," he was also repeatedy described as "endowed with might by Enlil, nourished with the sacred breastmilk of Ninharsag." The two kings belonged

Figure 99

to the First Dynasty of Lagash that was installed by Ninurta in reaction to the transfer of Kingship from Kish (that was under his aegis) to Uruk (under Inanna's patronage); and there are reasons to believe that all the nine kings of the first dynasty of Lagash were demigods in some manner.

The manner in which he was engendered, Eannatum claimed, allowed him to assume the title 'King of Kish', linking him—genealogically?—to the venerated Kish dynasty and its patron god Ninurta. While we can only guess how other kings of Kish qualified as demigods, no guesswork is required as Sumer's capital moved from Kish to Uruk; there **Utu is named as father** of the very first king, ***Mes.kiag.gasher***.

Utu (later known as *Shamash,* the 'Sun god'), it need be kept in mind, belonged to the second generation of great Anunnaki born on Earth, and his fathering of the head of a new dynasty must be considered a major milestone—a parenthood change from the Olden Gods who had come from Nibiru to an Earth born and bred male deity.

This generational change, with its genetic implications, was followed on the female side with ***Lugalbanda,*** the third king to reign in Uruk:

In his case, it was *a goddess*—**Inanna**—*who was identified as the mother;* the twin sister of Utu, she too was a second generation Anunnaki 'Earth Baby'. That was followed in Uruk by a second divine *maternal* involvement: The naming of the goddess **Ninsun** as wife of Lugalbanda and *her clear identification as the mother of their son* ***Gilgamesh***. And Ninsun—daughter of Ninurta and his spouse, Bau—was herself also an 'Earth Baby'.

A stone portrait of Ninsun found in Lagash with her name, **Nin.Sun** (pronounced 'Soon') clearly inscribed on it (Fig. 100), shows her dignified and serene; in fact, she was quite a master of court intrigues—in part, perhaps, out of necessity, being the mother of Lugalbanda's eleven children. A glimpse of her matchmaking is revealed in a segment of the Gilgamesh Epic, where she discussed with **Aya** (spouse of Utu) the selection of a young goddess as wife for Enkidu (as a reward for his undertaking to risk his life to protect Gilgamesh). Retaining much of

Figure 100

her parents' longevity (and the genes of their heroic stature), Ninsun lived long enough to mother several later kings. Her probable role in the life and death drama of the First Dynasty of Ur will be a highlight of our tale.

Sumer's capital remained in Ur for just over a century after the death of Gilgamesh, and then shifted to several other cities. Circa 2400 B.C. Ur served again, for the third time, as the national capital under an important king named ***Lugal.zagesi***. His many inscriptions included the claim that the goddess **Nisaba** was his mother:

> ***Dumu tu da dNisaba,***
> *Son born by/to divine Nisaba,*
> ***Pa.zi ku.a dNinharsag***
> *fed [with] holy milk by divine Ninharsag*

Nisaba, it will be recalled, was the astronomy goddess. In some texts she is called "sister of Ninurta," sharing with him Enlil as a father. But in the Great God List she was described as "divine Nisaba, a female, from the pure/sacred womb of divine Ninlil." In other words, she was an Earthborn daughter of Ninlil and Enlil, full sister of Nannar/Sin but only a half-sister of Ninurta (whose mother was Ninmah).

Here, then, in probable chronological order, is the picture that emerges from the nine kings of Kish, Lagash and Uruk whose demigod parentage has been verified:

> Etana: Of same seed as Adapa (= Enki's)
> Meskiaggasher: The god Utu is the father
> Enmerkar: The god Utu is the father
> Eannatum: Seed of Ninurta, Inanna put him on lap of Ninharsag for breastfeeding
> Entemena: Raised on Ninharsag's breastmilk
> Mesalim: "Beloved son" of Ninharsag
> Lugalbanda: Goddess Inanna his mother
> Gilgamesh: Goddess Ninsun is his mother
> Lugalzagesi: Goddess Nisaba his mother

These one-two-three punches reveal the significant post-Diluvial double shift in the affairs of gods and demigods: First, the 'Founding Fathers' progenitors who had come from Nibiru are replaced by the Earthborn generations. Then, through a stage involving the *'Sacred Breastmilk'*, the final change takes place: ***The female "Divine Womb" replaces the earlier male "Fecund Seed" and 'Pure Semen'.***

It is important to understand these changes, for they had long-term consequences. When the role of parenting demigods was taken over by the *Earthborn* gods and goddesses, was it just a matter of nature (i.e., getting old) taking its course, or did genealogical succession—through demigods—become more vital for those born on Earth *because their life cycles were shortened* by Earth, not Nibiru, being their home planet?

The records show that the Anunnaki did realize that those who had come and stayed on Earth (Enki, Enlil, Ninmaḫ) aged faster than those who stayed back on Nibiru; and that those who were born on Earth aged even faster. The changes from life on Nibiru to life on Earth apparently affected not only the longevity of the gods (and demigods), but also their physique, making them less giantlike as time went on. And then—we now know from advances in genetics—the switch of parenthood from the 'Fecund' Seed of the males to the female "Divine Womb" meant that the demigods from then on inherited both the general DNA as well as the specific Mitochondrial DNA of the female goddess.

These were changes whose significance will emerge as we follow the saga of gods and demigods to its concluding mystery.

In the biblical context, the crucial change in the realm of demigods from pre-Diluvial times can be summed up by us in a simple statement: Before, ***the sons of the gods*** "chose whichever they wanted from among the daughters of Man." Now ***the daughters of gods*** chose whichever they wanted from the sons of Men. The role of the goddesses in all that was epitomized by Sitar's six words. Where the mother was the deity, describing her as 'spouse' of the male no longer held true: *It was the male father who was chosen to be the companion of the goddess.* It was Inanna who said, "Come Gilgamesh, be thou my lover"; and with that, the Era of the Goddess had dawned.

* * *

Uruk's heroic age of Enmerkar, Lugalbanda and Gilgamesh petered out after the death of Gilgamesh. His son Ur.lugal and then grandson Utu. kalamma reigned a combined 45 years, and were followed by five more kings with a total throneship of 95 years. The King List deemed only one of them, ***Mes.h̲e,*** worthy of an extra word—noting that he was "a smith." All in all, according to the King List, "12 kings reigned (in Uruk) for 2,310 years; its kingship was carried to Ur."

The long reigns of the dynasties of what is now termed by scholars 'Kish I' and 'Uruk I' are recalled for their progress and stability, but not necessarily as peaceful times. On the national arena, as cities expanded to city-state size, disputes over boundaries, arable land, and water resources erupted into armed clashes. On the international stage, the hopes placed on the Inanna/Dumuzi union were dashed by Dumuzi's death and the ferocious war launched by Inanna against the accused Marduk. Of all the gods involved, the death of Dumuzi placed a tremendous emotional burden on Inanna; so much so that the ensuing events even led to her own death!

The tale is told in a text called *Inanna's Descent to the Lower World* (misleadingly titled by scholars *Inanna's Descent to the Netherworld*). It tells how Inanna, following the death of Dumuzi, went to the 'Lower World' domain of her sister Ereshkigal. The visit aroused Ereshkigal's suspicions, for not only did Inanna come uninvited, she also came to meet the god Nergal, her sister's spouse. So on Ereshkigal's orders Inanna was seized, killed with death rays, and her dead body was hung as a carcass . . .

When Inanna's handmaiden, who stayed back in Uruk, raised an alarm, the only one who could help was Enki. He fashioned two clay androids who could withstand the death rays, and activated them by giving one the Food of Life and the other the Water of Life. When they retrieved Inanna's lifeless body, "upon the corpse they directed the Pulser and the Emitter"; they sprinkled on her body the Water of Life and gave her the Plant of Life', "and Inanna arose."

Scholars have speculated that Inanna went to the Lower World to find Dumuzi's body; but in fact Inanna knew where the body was, because she had ordered for it to be mummified and preserved. She went, I have suggested in *Divine Encounters,* to seek from Nergal the fulfillment of a custom known from the Bible that required a brother (as Nergal was of Dumuzi) to sleep with the widow in order to obtain a son who will carry on the dead man's name; and Ereshkigal would have none of that.

Without doubt, these experiences profoudly affected Inanna's behavior and future actions; one of the notorius changes was the introduction by Inanna of the 'Sacred Marriage' rite, whereby a man of her choice (as often as not the king) had to spend with her the night on the anniversary of her unfulfilled wedding with Dumuzi; often, the man was found dead in the morning.

Thus, the transfer of the central capital to Ur was an attempt to gain respite by shifting responsibilities to Nannar/Sin—Ninurta's younger brother and Inanna's father.

* * *

Ur (pronounced 'Oor') was a new post-Diluvial city established as a 'cult center' for Enlil's son Nanna/Nannar (= 'The Bright One', an allusion to his celestial counterpart, the Moon). It was destined to play a major role in the affairs of gods and men, and its tale crossed paths with the biblical Abraham; but that was yet to take place when Ur would serve as Sumer's national capital for the third time. In the short span of what is called the 'Ur I' period, immediately following 'Uruk I', Ur—according to the King List—had four kings who reigned a total of 177 years; two of them are distinguished by their names—***Mes.Anne.pada*** and ***Mes.Kiag.nanna***.

Though Ur attained its most glorious—and tragic—time later on, in what is termed the 'Ur III' period, the archaeological evidence shows that the nearly two centuries of 'Ur I' were also times of high culture and great artistic and technological advancement. We know not whether it was cut short by mounting pressures on Sumer's borders by increasingly aggressive migrants, or by internal problems; the King List itself suggests that some turbulent events had taken place, causing the record-keepers to provide five (not four) royal names, amend one of them, and confuse reign lengths.

Whatever the troubling events might have been, the record shows that the national capital was abruptly moved from Ur to a minor city called Awan, and then in quick succession to cities called Hamazi and Adab, back (for a second time each) to Kish, Uruk, and Ur, shifted to cities called Mari and Akshak, then back again to Kish (III and IV)—all within a span of about two centuries.

Then, for the third time, the gods returned central kingship to Uruk, appointing as its king a strongman named ***Lugal.zagesi***. His mother, it will be recalled, was the goddess Nisaba, an aunt of Inanna, which (presumably) should have been enough to assure Inanna's blesssing. His first priority was to restore order among the quarreling and warring city-states, not refraining from use of his own troops to remove troublesome rulers. One of the cities subjected to punitive action by Lugal.zagesi was Umma—a city that served as 'cult center' for **Shara**, Inanna's son . . . So Lugalzagesi was gone soon after that, and the next King of Kings was a man of Inanna's own choice—a man who answered her call, "Come, be thou my lover!"

After all the millennia of gods in charge, a *goddess* was now in full command.

'HERO' BY ANY NAME

Two of the names of Ur I—***Mes.anne.pada*** and ***Mes.kiag.nanna***—are noteworthy because, as that of Uruk's ***Mes.he*** (***He*** = 'Fullness/Plenty'), they have as a prefix the syllable-word **Mes** that we have encountered before—in ***Mes.Kiag.gasher***, the very first king of Kish whose father was the god Utu, and in the later king of Kish, ***Mes .Alim*** (***Alim*** = 'Ram'), who claimed to have been the "beloved son" of Ninharsag.

This raises the question: Did **Mes** as a prefix (or ***Mesh*** a suffix, as in Gilgamesh) identify the person as a demigod? Apparently so, because the term **Mes** in fact meant 'Hero' in Sumerian—the very meaning of the Hebrew term *Gibbor* used in Genesis 6 to define the demigods!

Such a conclusion is supported by the fact that an Akkadian text catalogued BM 56488 concerning a certain temple contains the statement:

> *Bit sha dMesannepada ipushu*
> Temple which divine Mesannepada built
> *Nanna laquit ziri ultalpit*
> Nannar, the seed giver, destroyed

—a statement that both assigns the determinative 'divine' to Mesannepada, and, by referring to the god Nannar/Sin as "the seed giver," indicates which god was the procreator of this demigod.

One must also wonder, in view of other meaning similarities that we have already mentioned, whether the Sumerian **Mes** and the Egyptian *Mes/Mses* as in Thoth*mes* or Ra*mses* (meaning "issue of" in Pharaonic claims of divine parentage) do not stem from some common early source.

Our conclusion that royal Sumerian names starting (or ending) with **Mes** indicate demigod status will serve as a clue to unlock varied enigmas.

XIV
Glory of Empire, Winds of Doom

One day my queen,
After crossing heaven, crossing Earth—
Inanna—
After crossing heaven, crossing Earth—
After crossing Elam and Shubur,
After crossing [. . .],
The hierodule approached weary, fell asleep.
I saw her from the edge of my garden.
I kissed her, copulated with her.

So did a gardener later known as Sharru-kin ('Sargon' in English) describe his chance encounter with the goddess Inanna. Since the goddess, weary from her flying about, was asleep, one cannot say that it was a case of 'Love at first sight'; but from what ensued it is obvious that Inanna liked the man and his lovemaking. Inanna's invitation to him to her bed, with the throne of Sumer thrown in, lasted 54 years: "While I was a gardener, Ishtar granted me her love; for four and fifty years I exercised kingship; the Black-Headed People I ruled and governed," Sargon wrote in his autobiography.

How did Inanna persuade the Anunnaki leadership to entrust Sumer and its people—here called by their nickname ***Sag.ge.ga,*** the Black-Headed Ones—to the man whose kiss changed history, is nowhere made clear. His name-title *Sharru-kin* (= 'Truthful Ruler') was not Sumerian; it was in the 'Semitic' tongue of the ***Amurro,*** the 'Westerners', of the 'Semitic' speaking region northwest of Sumer; and his features, preserved in a bronze sculpture (Fig. 101), confirm his non-Sumerian extraction. The brand-new capital city built for him, ***Agade,*** was better known by its 'Semitic' name *Akkad*—from which the term *Akkadian* for the language.

The Sumerian King List, recognizing the significance of this king, provides the information that from Uruk under Lugal.zagesi "kingship to Agade was carried" and notes that Sharru.kin, "a date-grower and cupbearer of Ur.zababa," built Agade and reigned there for 56 years.

The position of Cup Bearer was one of high rank and great trust, usually held by a prince, in royal courts not only in Mesopotamia, Egypt, and elsewhere in the ancient world—it was so, it will be recalled, even on Nibiru (where Anu served as Alalu's cupbearer). Indeed, some

Figure 101

of the earliest Sumerian depictions that scholars call 'Libation scenes' might be depictions of the king (bare naked to show total subservience) acting as cupbearer for the deity (see Fig. 77).

Urzababa was a king in Kish, and the statement implies that Sargon was a royal prince there. Yet Sargon himself, in the autobiographical text known as *The Legend of Sargon,* chose to wrap his origin in mystery:

> Sargon, the mighty king of Agade, am I.
> My mother was a high priestess;
> I knew not my father.
> My mother, the high priestess who conceived me,
> in secret she bore me.

Then, as in the story of the birth of Moses in Egypt a thousand years later, Sargon continued:

> She set me in a basket of rushes,
> with bitumen sealed the lid.
> She cast me into the river, it did not sink me.
> The river bore me, carried me to Akki the gardener.
> Akki the irrigator lifted me up when he drew water.
> Akki the irrigator as his son made me and reared me.
> Akki, the irrigator, appointed me as his gardener.

The explanation for Sargon's odd avoidance of claiming princehood might be found in the fact that in time Sargon's own daughter Enheduanna served as high-priestess-cum-hierodule in the temple of the god Nannar/Sin in Ur—a position deemed one of great honor. By claiming the same position for his mother, Sargon left open the possibility that his 'unknown father' might have been a god—which would make him, Sargon, a demigod.

It is quite possible that Sargon's Amorite ancestry might have been a favorable consideration, in view of the pressures on Sumer by migrants from the west and northwest. The same thinking, of making adversaries part of the family, probably led to the decision to establish a new, neutral national capital whose name meant 'Union'; its location marked the addition of territories called Akkad, north of olden Sumer, to create a new geopolitical entity called 'Sumer & Akkad'; and henceforth, Inanna was widely known by her Akkadian name ***Ishtar***.

Circa 2360 B.C., Sargon set out from that new capital to establish law and order, starting with the defeat of Lugal.zagesi (who, the reader will recall, dared attack the city of Ishtar's son, the god Shara). Bringing one olden city after another under his control, he turned his prowess against neighboring lands. To quote from a text known as *The Sargon Chronicle,* "Sharru-kin, king of Agade, rose to power in the era of Ishtar. He had neither rival nor opponent. He spread his terror-inspiring glance over all the countries. He crossed the sea in the east; he conquered the country in the west in its full extent."

For the first time since its inception millennia earlier, the whole First Region was firmly ruled from a national capital, from the Upper Sea (the Mediterranean) to the Lower Sea (the 'Sea in the East', the Persian Gulf); in that, it was the first historically known empire—and quite an empire it was: Inscriptions and archaeological evidence confirm that Sargon's dominion extended to the Mediterranean coast in the west, the Khabur River in Asia Minor in the north, lands in the northeast that were to become later on Assyria, and sites on the eastern coast of the Persian Gulf. And though Sargon acknowledged (when necessary) the authority of Enlil, Ninurta, Adad, Nannar, and Utu, his conquests were carefully carried out "by the order of my mistress, the divine Ishtar." It was indeed, as the inscriptions said, the ***Era of Ishtar***.

As an imperial capital, Agade was a grandeur to see. "In those days," a Sumrian text reported, Agade was filled with riches of precious metals, of copper and lead and slabs of lapis lazuli. "Its granaries bulged at the sides, its old men were endowed with wisdom, its old women were endowed with eloquence, its young men were endowed with the strength of weapons. Its little children were endowed with joyous hearts . . . The city was full of music." A grand new temple for Sitar made clear which deity held sway over all of that: "In Agade," a Sumerian historiographic text stated, "did holy Inanna erect a temple as her abode; in the Glittering Temple she set up a throne." It was the crown jewel of shrines to her that had to be erected in virtually every Sumerian city, outshining even the sacred Eanna in Uruk; and that was a mistake.

Sargon too, growing haughty and overambitious, began to commit grave errors, including sending his troops into cities beholden to Ninurta and Adad. And then he committed a fateful act: desecrating Babylon. The territory designated 'Akkad', north of olden Sumer, included the site of Babylon, the very place where Marduk, seeking supremacy, had attempted to build his own launch tower (the Tower of Babel incident). Now Sargon "took away soil from the foundations of Babylon, and built upon the soil another *Bab-ili* near Agade."

To understand the severity of this unauthorized act, one needs recall that *Bab-ili* (as 'Babylon' was called in Akkadian) meant 'Gateway of the Gods', a sanctified place; and that Marduk was persuaded to give up his attempt on condition that the site will be left undisturbed, as 'hallowed ground'. Now Sargon "took away soil from the foundations of Babylon" to use as foundation soil for another Gateway of the Gods, adjoining Agade. The sacrilege naturally enraged Marduk, and reignited the clan conflicts. But Sargon not only broke the taboo regarding Babylon—he also planned to create at Agade his (or Inanna's?) own 'Gateway of the Gods'; and that angered Enlil.

* * *

The resulting prompt removal (and death) of Sargon did not end the 'Era of Ishtar'. With the consent of Enlil, she placed Sargon's son Rimush on the throne in Agade; but he was replaced after a brief nine years by his brother Manishtushu, who lasted fifteen years. Then ***Naram-Sin,*** the son of Manishtushu, ascended the throne—and once more, Inanna/Ishtar had as king a man to her heart's desire.

Naram-Sin, whose theophoric name meant "Whom [the god] Sin loves," used the Akkadian name *Sin* of Inanna's father rather than the Sumerian **Nannar**. Capably building on the imperial foundations attained by his grandfather, he combined military expeditions with the expansion of commerce, sponsoring trading posts by Sumerian merchants in far-flung places and creating trade routes on an international scale, reaching as far north as the boundary of the Hittite domain of **Ishkur**/Adad, Nannar's brother.

Naram-Sin's two-track policy of stick and carrot, however, failed to counter the rising number of cities, especially in the west, siding with Marduk's renewed ambitions for supremacy. Highlighting the fact that his spouse, ***Sarpanit,*** was an Earthling and that his Earthborn son Nabu also married one (named ***Tashmetum***), Marduk was gaining adherents among the masses. In Egypt, where Marduk/Ra has been worshipped as the hidden Amen/Amon, the expectations for Marduk's ultimate victory were reaching messianic fervor, and Egyptian Pharaohs began to thrust northward, to seize control of the Mediterranean's coastal lands.

It was thus that, with the blessing and guidance of Inanna/Ishtar, Naram-Sin launched against the "sinning cities" in the west what was by that time the greatest military expedition ever. Capturing what was later known as Canaan, he kept advancing all the way south to Magan (ancient Egypt). There, his inscriptions state, "he personally caught the king of Magan." His merciless advance and capture of adversary kings was commemorated on a stone plaque that showed a glowing Ishtar offering him a victory wreath (Fig. 102). Having entered and crossed

Figure 102

Figure 103

the forbidden Fourth Region with its Spaceport, Naram-Sin haughtily depicted himself on a victory stela (Fig. 103) godlike astride a rocketship to the heavens. He then went to Nippur to demand that Enlil endorse him as "King of the Four Regions." Enlil was not there; so "Like a hero accustomed to high handedness he put a restraining hand on the Ekur," Enlil's sacred precinct.

These were unprecedented acts of disobedience and sacrilege; Enlil's reaction is detailed in a text known as *The Curse of Agade.* He summoned the Anunnaki leadership to an Assembly; all the great gods, Enki included, attended—but Inanna did not show up. Ensconed in the venerated Eanna temple in Uruk, she sent back defying words, demanding that the gods declare her "Great Queen of Queens"—the supreme female deity.

"The heavenly Kingship was seized ***by a female!***" the ancient text noted in astonishment; "Inanna changed the rules of Holy Anu!"

In their Assembly, the gods' decision was made—to put an end to all that by wiping Agade off the face of the Earth. Troops loyal to Ninurta from Gutium, a land across the Zagros Mountains, were brought in, and they systematically destroyed Agade to oblivion. The gods decreed that its remains shall never be found; and indeed, to this day, no one is even certain where exactly Agade had been located. With the death of the city, Naram-Sin too was gone from the records.

As to Inanna/Ishtar, her father Nannar/Sin fetched her from Uruk to Ur. "Her mother Ningal greeted her back at the temple's entrance. 'Enough, enough innovations' to Inanna she said," according to the texts; her home was to be with Nannar's family in the sacred precinct of Ur.

By 2255 B.C., the 'Era of Ishtar' was over. But the empire which she had brought about—as well as the challenges to olden authority—left their permanent mark on the ancient Near East.

* * *

For about a century thereafter, there was no kingship in a national capital in Sumer and Akkad. "Who was king? Who was not king?" the Sumerian King List itself noted as a way of describing the situation. *De facto* the country was administered by Ninurta from his 'cult center' in Lagash—a city whose written records, artifacts, and sculptures have served as a major source of information about Sumer, the Sumerians, and the Sumerian civilization.

Archaeological and documentary evidence from the site (now called

Tello) shows that circa 2600/2500 B.C.—about three centuries before Sargon of Akkad—dynastic rule began in Lagash with a ruler named Lugal.shu.engur; that first dynasty included such famed demigod heroes as Eannatum (of artificial insemination fame). Dynastic rule continued in Lagash uninterrupted for more than half a millennium, indicating outstanding stability throughout turbulent times; the list of its kings runs to 43 names!

The kings of Lagash, who preferred the title ***Patesi*** (= 'Governor') to that of ***Lugal,*** left behind countless votive and other inscriptions. To judge by the textual evidence, those were enlightened kings who strived to shape the people's lives in accordance with their god's high standards of justice and morality; the greatest honor a king could attain was to be granted by Ninurta the epithet 'Righteous Shepherd'. A king named ***Urukagina*** instituted, some 4,500 years ago, a code of laws that prohibited the abuse of official powers, the "taking away" of a widow's donkey, or the delay by a supervisor of the wages of daily workers. Public works, such as canals for irrigation and transportation, and communal buildings, were deemed a personal duty of the king. Festivals that involved the whole populace, such as the Festival of First Fruits, were introduced; literacy, evidenced by some of the most perfect cuneiform script, was encouraged; and some of the finest Sumerian sculptures—two thousand years before classic Greece!—come from Lagash (see Figs. 31, 33).

Yet none of the Lagash rulers are mentioned in the Sumerian King List, and Lagash had never served as a national capital. Once the seat of national Kingship was transferred from Kish to Uruk—in religio-political terms, from the aegis of Ninurta to the dominance of Inanna—what Ninurta did was to establish his own redoubt, protected by what were then the best trained troops in the land, outside the reach of Inanna's whims and ambitions. So it was from Lagash that Ninurta restored Enlilite authority and brought about a century of respite to Sumer after the Inanna/Naram-Sin upheavals; but it was a shrinking Sumer & Akkad, subjected to relentless pressures as Marduk continued to seek Supremacy on Earth.

It was to counteract those ambitions that circa 2160 B.C. Enlil authorized Ninurta to erect, in Lagash, an astonishing and unique new temple that would declare *Ninurta's* claim to supremacy. To make it clear, the temple was to be called ***E.Ninnu***—"House/Temple of Fifty," affirming Ninurta as the 'next Enlil' with the Rank of Fifty, just below Anu's 60.

Some of the most extensive inscriptions found in the excavated remains of Lagash—with amazing details that could come out of a *Twilight Zone* TV episode—concern the building of that new temple in the Girsu (the sacred precinct of Lagash) by a king named ***Gudea*** (= 'The Anointed One'). The story, which is recorded on clay cylinders that are now on display in the Louvre museum in Paris, began with a dream that Gudea had. In the dream, "a man, bright and shining like heaven . . . who wore the headdress of a god" appeared and commanded Gudea to build him a temple. A female, "a woman carrying the structure of a temple on her head," appeared next; holding a tablet with a celestial map, she pointed at a particular star. Then a second male deity appeared, holding in one hand a tablet with a design on it and in the other hand a building brick.

Awakened, Gudea was astounded to discover the tablet with the design lying on his lap, and the building brick in a basket by his side! Completely baffled by the experience (commemorated by Gudea in one of his statues, Fig. 104), Gudea journeyed to the "House of Fate-solving," abode of the goddess **Nina** in her cult center Sirara, and asked her to solve the dream and the meaning of the out-of-nowhere objects.

The first god, Nina said, was **Nin.girsu** (= 'Lord of the Girsu', alias Ninurta); "for thee to build a new temple he commands." The goddess is **Nisaba**; "to build the temple in accordance with the Holy Planet she instructs thee." The other god is **Ningishzidda**; the sacred brick he gave you, is to be used as a mold; the carrying-basket means that you have been assigned the task of construction; the tablet with the design on it is the architectural plan of the seven-stage temple; its name, she said, shall be ***E.Ninnu***.

With most other kings just proud to engage in repairing existing temples, the choosing of Gudea to build a brand new one from

Figure 104

foundations up was an unusual honor. With joy he set out to build it, mobilizing the whole populace for the project. The architectural requirements, he found out, were far from simple; there was to be at the top *a domed observatory*—"shaped like the vault of heaven"—to determine star and planetary positions after nightfall, and in the forecourt two stone circles were to be erected to determine constellations at *the moment of sunrise on Equinox Day*. There was also need to construct two special sunken enclosures, one for Ninurta's aircraft, the "Divine Black Bird," and the other for his "Awesome Weapon." In his clearly written inscriptions in perfect Sumerian script (example, Fig. 105) Gudea states that he had to go back repeatedly to the deities for guidance, and "had no good sleep until it was completed." At one point he was ready to give up, but in a "command-vision" was ordered "the building of the Lord's House, the Eninnu, to complete."

The preliminary events and details of the complex construction are inscribed on what is called *Gudea Cylinder A*. 'Cylinder B' is devoted to the elaborate rites connected with the temple's inauguration—precisely on

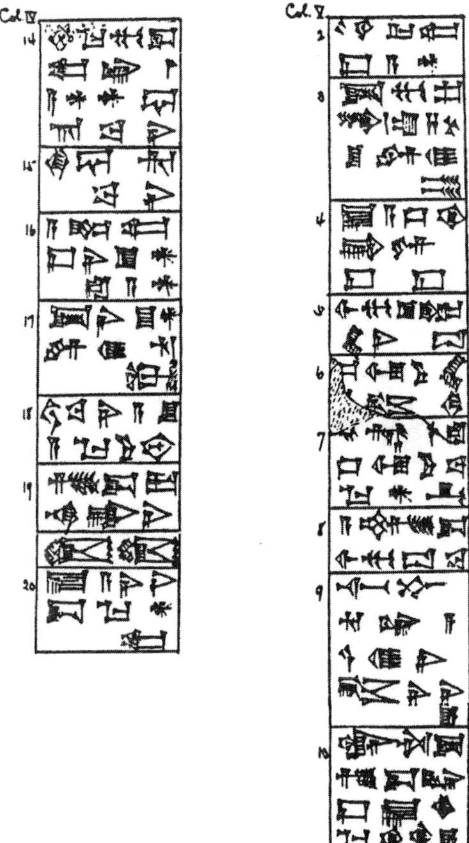

Figure 105

New Year's Day—and the ceremonies attendant on the arrival of Ningirsu and Bau at the Girsu and their entry into their new temple-home. It ends with a recorded blessing of Gudea by Bau in gratitude for his construction efforts; his reward was ***Nam.ti muna.sud***—"His Lifetime Sustained/Prolonged" (without an explanation how it was granted).

Introducing himself in Cylinder A, Gudea stated that the goddess **Nina**—a daughter of Enlil and Ninlil, a half-sister of Ninurta—was his mother, repeatedly calling her "my mother" in Cylinder A; and in the blessing by Bau at the end of Cylinder B, she twice referred to him as "son of Nina." These texts also shed light on the manner of his birth: The goddess Nina brought him forth from *seed implanted in her womb* *by the goddess Bau:* "The germ of me thou didst receive within thyself, in a sacred place thou didst bring me forth," he said to Nina; he was "a child by Bau brought forth."

Gudea, in other words, asserted that he was a demigod, engendered by Bau and Nina of the Enlil/Ninurta clan.

* * *

The challenge posed to Marduk by the Eninnu temple was compounded by the roles of the deities Ningishzidda and Nisaba—both known and worshipped in Egypt: The former as the god Thoth and the latter as the goddess Sesheta. The active participation of Ningishzidda/Thoth in the project was especially significant, since he was a son of Enki/Ptah and a half-brother of Marduk/Ra, with whom he had repeatedly quarreled. That was not the only inner rift with Marduk: His other half-brother, Nergal (spouse of Enlil's granddaughter Ereshkigal), also sided frequently with the Enlilites.

Yet all that failed to stop Marduk and Nabu from gaining adherents and territorial control. The growing problem for the Enlilites was the fact that Ninurta, the presumptive heir to Enlil and Anu, had come from Nibiru—whereas Marduk and Nabu had Earthling affinities. In desperation, the Enlilites dropped the 'Ninurta Strategy' and switched to a 'Sin Tactic', transferring the seat of national Kingship to Ur—the 'cult center' of Nannar, *an Earthborn son of Enlil,* who unlike Ninurta also had an Akkadian name: *Sin.*

Ur, situated between Eridu to the south and Uruk in the north along the Euphrates River, was by then Sumer's thriving commercial and manufacturing center; its very name, which meant "urban, domesticated place," came to mean not just 'city' but '*The* City' and spelled prosperity and well-being. Its gods (see Fig. 97) Nannar/Sin and his spouse **Ningal** (*Nikkal* in Akkadian) were greatly beloved by the people of Sumer; unlike other Enlilites, Nannar/Sin was not a combatant in the wars of the gods. His selection was meant to signal to people everywhere, even in the "rebel lands," that under his leadership an era of peace and prosperity will begin.

In Ur the deities' temple-abode was a great ziggurat that rose in stages within a walled sacred precinct, where a variety of structures housed priests, officials, and servants. One of the buildings within the walled section was the ***Gipar*** (= 'Nighttime Abode') within which was the ***Gigunu,*** the 'Chamber of nighttime pleasures' for the god; for though Nannar/Sin was monogamous and had only one spouse (Ningal), he could (and did) enjoy in the Gipar the company of hierodules ('Pleasure Priestesses') as well as concubines (by whom he could have children).

Beyond those walls there extended a magnificent city with two harbors and canals linking it to the Euphrates River (Fig. 106), a great city

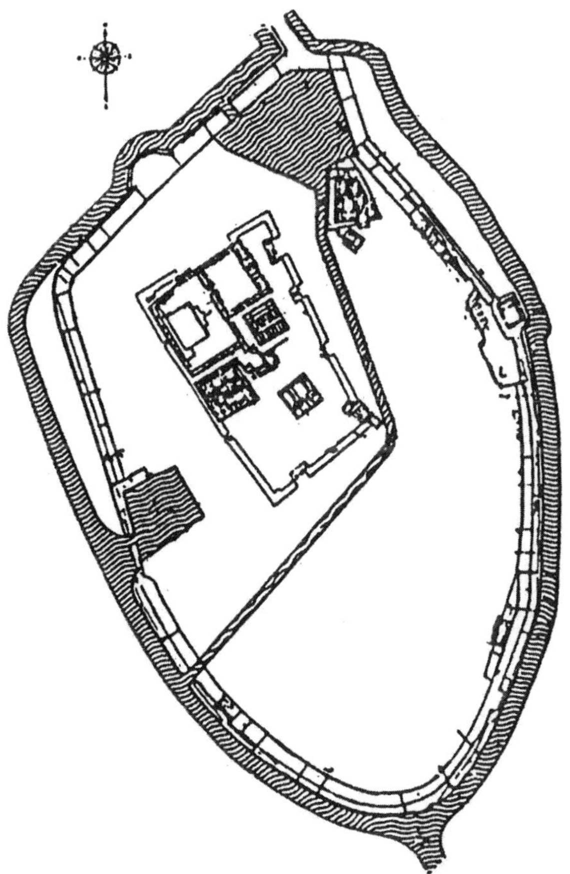

Figure 106

with the king's palace, administrative buildings, lofty gates, avenues for promenading, a public square for festivals, a marketplace, multilevel private dwellings (many two-storied), schools, workshops, merchants' warehouses, and animal stalls. The imposing ziggurat with its monumental stairways (see Fig. 35), though long in ruins, dominates the landscape to this day, even after more than 4,000 years.

(Ur, let it be noted, was the 'Ur of the Chaldees' in which the biblical story of Abraham the Hebrew began, the starting point of his migration to Harran and then to Canaan. Born in Nippur, ***Abram*** grew up in Ur where his father served as a *Tirhu,* an Omen Priest skilled in astronomy. How his tale and mission interlocked with the events and fate of Sumer, has been told by us in detail in *The Wars of Gods and Men.*)

To restart afresh Kingship in and from Sumer, the choice of a new king was also carefully made. The new king, named ***Ur-Nammu*** (= 'The joy of Ur'), was selected by Enlil and approved by Anu; and he was no mere Earthling—*he was a demigod.* Born in Uruk, he was a son—"the beloved son"—of the goddess **Ninsun** (who had been the mother of Gilgamesh)—a birth (according to the inscriptions) approved by Anu and Enlil and witnessed by Nannar/Sin. Since this divine genealogy (including the claim that Ninharsag helped raise him) was restated in numerous inscriptions during Ur-Nammu's reign, in the presence of Nannar and other gods, one must assume that the claim was factual. It was a claim that placed Ur-Nammu in the very same status as that of Gilgamesh, whose exploits were well remembered and whose name remained revered. The choice was thus a signal, to friends and foes alike, that the glorious days under the unchallenged authority of Enlil and his clan were back.

The inscriptions, the monuments, and the archaeological evidence attest that Ur-Nammu's reign witnessed extensive public works, restoration of river navigation, and the rebuilding and protection of the country's highways. There was a surge in arts, crafts, schools, and other improvements in social and economic life. Enlil and Ninlil were honored with renovated and magnified temples; and for the first time in

Sumer's history, the priesthood of Nippur was combined with that of Ur, leading to a religious revival. (It was at that time, by our calculations, that the Omen Priest ***Terah̲***, Abram's father, was transferred from Nippur to Ur.)

Treaties with neighboring rulers to the east and northeast spread the prosperity and well-being; but the enmity stirred up by Marduk and Nabu in the west was rising. The situation in the "rebel lands" and "sinning cities" bordering the Mediterranean Sea demanded action, and in 2096 B.C., Ur-Nammu launched a military campaign against them. But as great a builder and economic 'shepherd' as he was, he failed as a military leader: In the midst of battle his chariot got stuck in the mud; Ur-Nammu fell off it and was "crushed like a jug." The tragedy was compounded when the boat returning Ur-Nammu's body to Sumer "in an unknown place had sunk; the waves sank it down, with him on board."

When news of the defeat and the tragic death of Ur-Nammu reached Ur, a great lament went up. The people could not understand how such a religiously devout king, a righteous shepherd—a demigod!—could perish so ingeniously. "Why did the Lord Nana not hold him by the hand?" they asked; "Why did Inanna, Lady of Heaven, not put her noble arm around his head? Why did the valiant Utu not assist him?" There could be only one plausible explanation, the people of Ur and Sumer concluded: "Enlil deceitfully changed his decree"—these great gods went back on their word; and faith in them was profoundly shaken.

It was probably not by chance that exactly upon the shocking death of Ur-Nammu in 2096 B.C. Abram's father moved his family from Ur to ***Harran*** (= 'The Caravanry'), a major city at what was then Sumer's link with the Land of the Hittites. Situated at the headwaters of the Euphrates River and located at the crossroads of international trade and military land and river routes, Harran was surrounded by fertile meadows perfect for sheepherding. It was founded and settled by merchants from Ur who came there for its local sheep's wool, skins, and leather and imported metals and rare stones, and brought in exchange Ur's famed woolen garments and carpets. The city also boasted the second largest temple to Nannar/Sin after Ur and was often called "the second Ur."

Ur-Nammu's ascent to the throne in Ur in 2113 B.C. ushered a period known as 'Ur III'. ***It was Sumer's most glorious period, and the timeslot in which Monotheism—the belief in one universal creator God—had its roots***

It was also Sumer's most tragic period, for before that century was over, Sumer was no more.

* * *

Following Ur-Nammu's tragic death, the throne of Ur was taken over by his son Shulgi. Eager to claim the status of a demigod as that of his father, he asserted in his inscriptions that he was born under divine auspices: The god Nannar himself arranged for the child to be conceived in Enlil's temple in Nippur through a union between Ur-Nammu and Enlil's high priestess, so that "a Little Enlil, a child suitable for kingship and throne, shall be conceived." He got into the habit of calling the goddess Ningal, Nannar's spouse, "my mother" and Utu/Shamash (their son) "my brother." He then asserted in self-laudatory hymns that "a son born of Ninsun am I" (though in another hymn he was her son only by adoption). These different and contradicting versions cast doubt on the validity of his claims to demigodship.

The royal annals indicate that soon after he had ascended the throne, Shulgi launched an expedition to the outlying provinces, including the 'rebel lands'; but his 'weapons' were offers of trade, peace, and his daughters in marriage. His route embraced the two destinations of the still revered Gilgamesh: The Sinai peninsula (where the Spaceport was) in the south and the Landing Place in the north, observing however the sanctity of the Fourth Region by not entering it. On the way, he paused to worship at the "Place of Bright Oracles"—the place we know as Jerusalem. Having thus venerated the three space-related sites, he followed the 'Fertile Crescent'—the arching trade and migration east-west route dictated by geography and water sources—and returned to Sumer.

When Shulgi came back to Ur, he was granted by the gods the title

'High Priest of Anu, Priest of Nannar'. He was befriended by Utu/Shamash; and then was given the 'personal attention' of Inanna/Ishtar (whose abode has been in Ur since the demise of Naram-Sin). Shulgi's 'Peace Offensive' bore fruit for a while, leading him to turn from affairs of state to become Inanna's lover. In numerous love songs that have been found in the ruins of Ur, he boasted that Inanna "granted me her vulva in her temple."

But as Shulgi neglected affairs of state to indulge in personal pleasures, the unrest in the 'rebel lands' grew again. Unprepared for military action, Shulgi relied on Elamite troops to do the fighting, and started to build a fortified wall to protect Sumer against foreign incursions. It was called the 'Great West Wall', and scholars believe that it ran from the Euphrates to the Tigris Rivers north of where Baghdad is situated nowadays. An unintended result of that was that the heartland of Sumer was cut off from the provinces in the north. In 2048 B.C. the gods, led by Enlil, had enough of Shulgi's state failures and personal *dolce vita,* and decreed for him "the death of a sinner." Significantly, it was exactly then that, by divine order, Abram left Harran for Canaan . . .

Also in that same year, 2048 B.C., Marduk arrived in Harran, making it his headquarters for the next 24 years. His arrival there, recorded in a well-preserved clay tablet (Fig. 107), posed a new and direct challenge to Enlilite hegemony. Besides the military significance, the move deprived Sumer of its economically vital commercial ties. A shrunken Sumer was now under siege.

Marduk's chess move to establish his command post in Harran enabled Nabu "to marshal his cities, toward the Great Sea to set his course." Individual site names reveal that those places included the all-important Landing Place in Lebanon and the Mission Control city of Shalem (alias Jerusalem). And then came Marduk's claim that the Spaceport Region was no longer neutral—it was to be considered a Marduk and Nabu domain. With Egypt his original dominion, he now controlled all the space-related facilities.

The Enlilites, understandably, could not accept such a situation. Shulgi's successor, his son ***Amar-Sin,*** lost no time launching one

Figure 107

military expedition after another, culminating with an ambitious and notable expedition to punish the 'Rebel Lands of the West' (the biblical Canaan). And so it was that in the 7th year of his reign, in 2041 BC., Amar-Sin led a great military alliance against the "sinning cities" in the west (including Sodom and Gomorrah), hoping to regain control of the Spaceport; he was, I have suggested in *The Wars of Gods and Men,* the 'Amarphel' of Genesis 14.

The clash is recorded in the Bible as the War of the Kings of the East against the Kings of the West. In that first Great International War of antiquity, Abram was a participant: Commanding a cavalry of camel riders called *Ish Nar*—a literal Hebrew rendering of the

Sumerian ***Lu.nar*** (= 'Cavalryman')—he successfully prevented the invaders from reaching the Spaceport (map, Fig. 108). He then pursued the retreating invaders all the way to Damascus (nowadays Syria), to rescue his nephew Lot whom they had taken captive in Sodom. The conflict

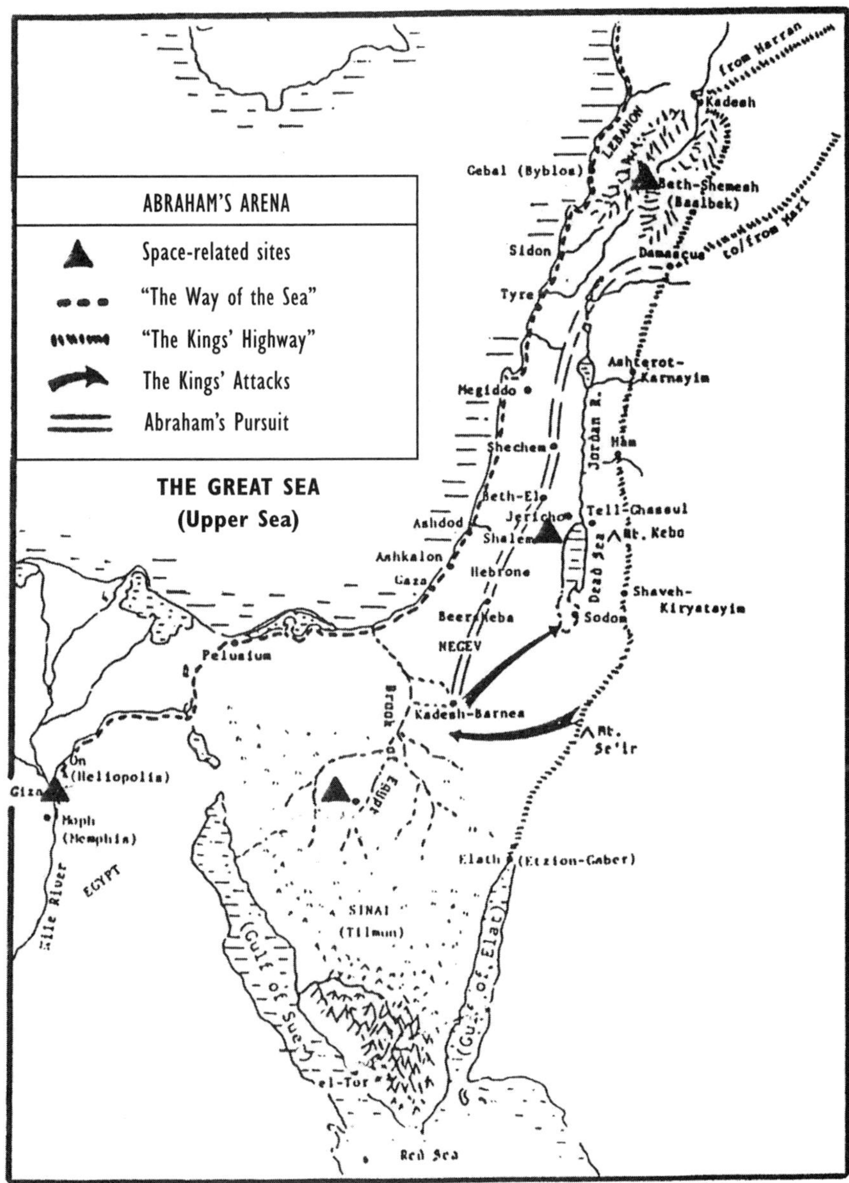

Figure 108

between the gods was clearly becoming a far-flung multi-nations war.

Amar-Sin died in 2039 B.C.—felled not by an enemy lance but by a scorpion's bite. He was replaced by his brother Shu-Sin; the data for his nine years' reign record two military forays northward but none westward; they speak mostly of his defensive measures. He relied mainly on building new sections of the Wall of the West; the defenses, however, were moved each time ever closer to Sumer's heartland, and the territory controlled from Ur kept shrinking.

By the time the next (and last) 'Ur III' king, Ibbi-Sin, ascended the throne in 2029 B.C., invaders from the west broke through the defensive Wall and were clashing with Ur's 'Foreign Legion', Elamite troops, in Sumerian territory. Directing and prompting the Westerners was Nabu. His divine father, Marduk himself, was waiting in Harran for the recapture of Babylon.

To the old reasons for seeking Supremacy (starting with his father Enki having been deprived of the succession rights), Marduk now added a 'Celestial' argument, claiming that his time for Supremacy had come because Enlil's zodiacal Age of the Bull ('Taurus') was ending, and his era, the Age of the Ram ('Aries'), was dawning. Ironically, it was his own two brothers who pointed out that astronomically observed, the zodiacal constellation of the Ram had not yet begun: Ningishzidda said so from the observatory in Lagash, and Nergal from the scientific station in the Lower World. But his brothers' findings only angered Marduk and intensified Nabu's recruiting of fighters for Marduk.

Frustrated and desperate, Enlil convened the great gods to an emergency assembly; ***it approved extraordinary steps that changed the future forever.***

* * *

Amazingly, various written records from antiquity have survived, providing us not just with an outline of events but with great details about the battles, the strategies, the discussions, the arguments, the participants and their moves, and the crucial decisions that resulted in the most profound upheaval on Earth since the Deluge.

Augmented by the Date Formulas and varied other references, the principal sources for reconstructing those dramatic events are the relevant chapters in Genesis; Marduk's statements in a text known as *The Marduk Prophecy;* a group of tablets in the 'Spartoli Collection' in the British Museum known as *The Khedorla'omer Texts;* and a long historical/autobiographical text dictated by the god Nergal to a trusted scribe, a text known as the *Erra Epos.* As in a movie—usually a crime thriller—in which the various eye witnesses and principals describe the same event not exactly the same way, but from which the real story emerges, so are we able to retrieve the actual facts in this case.

Marduk, we learn from those sources, did not personally attend the emergency council summoned by Enlil, but sent to them an appeal in which he repeatedly asked: "Until when?" The year, 2024 B.C., marked the 72nd anniversary of his life on the run—the time it takes the zodiacal circle to move one degree. It was 24 years since he had been waiting in Harran; and he asked: "Until When? When will my days of wandering be completed?"

Called to make the Enlilite case, Ninurta blamed everything on Marduk, even accusing his followers of defiling Enlil's temple in Nippur. Nannar/Sin's accusations were mainly against Nabu. Nabu was summoned, and "Before the gods the son of his father came." Speaking for his father, he blamed Ninurta; voicing accusations against Nergal, he got into a shouting match with Nergal (who was present); and "showing disrespect, to Enlil evil he spoke," accusing the Lord of the Command of injustice and of condoning destruction. Enki spoke up: "What are Marduk and Nabu actually accused of?" he asked. His ire was directed especially at his son Nergal: "Why do you continue the opposition?" he asked him. The two argued so much that in the end Enki shouted to Nergal to get out of his presence.

It was then that Nergal—vilified by Marduk and Nabu, ordered out by his father, Enki—"consulting with himself," concocted the idea of resort to the "*Awesome Weapons.*"

He did not know where they were hidden, but knew that they

existed on Earth, locked away in a secret underground place (according to a text catalogued as CT-xvi lines 44–46) somewhere in Africa, in the domain of his brother Gibil. Based on our current level of technology, they can be described as ***seven nuclear devices:*** "Clad with terror, with a brilliance they rush forth." They were brought to Earth unintentionally from Nibiru by the fleeing Alalu, and were hidden away in a secret safe place a long time ago; Enki knew where; so did Enlil.

Meeting again as a War Council, the gods, overruling Enki, voted to follow Nergal's suggestion to give Marduk a punishing blow. There was constant communication with Anu: "Anu to Earth the words was speaking, Earth to Anu the words pronounced." He made clear that his approval for the unprecedented step of using the "Awesome Weapons" was limited to depriving Marduk of the Sinai Spaceport, but that neither gods nor people should be harmed: "Anu, lord of the gods, on the Earth had pity," the ancient records state. Choosing Nergal and Ninurta to carry out the mission, the gods made absolutely clear to them its limited and conditional scope.

In 2024 B.C. Ninurta (called in the epic* Ishum, *'The Scorcher') and Nergal (called in the epic* Erra, *'The Annihilator') unleashed nuclear weapons that obliterated the Spaceport and the adjoining "sinning cities" in the plain south of the Dead Sea.

Abraham, according to the Bible, who was then encamped in the mountains overlooking the Dead Sea, was visited earlier that day by three *Malachim* (translated 'angels' but literally meaning 'emissaries') and was forewarned by their leader of what was about to happen. The other two went ahead to Sodom, where Abraham's nephew Lot dwelt. That night, we know from the *Erra Epos,* Ishum/Ninurta "to the Mount Most Supreme set his course" in his Divine Black Bird. Arriving there,

> He raised his hand (and)
> the mount was smashed.
> The plain by the Mount Most Supreme

he then obliterated;
in its forests not a tree stem was left standing.

With two pinpointed nuclear drops, the Spaceport was obliterated by Ninurta—first the 'Mount Most Supreme' ('Mount Mashu' of the Gilgamesh Epic) with its inner tunnels and hidden facilities; then the adjoining plain that served for landing and takeoff. The scar in the Sinai Peninsula is still visible to this day, as a NASA photograph from space shows (Fig. 109); the plain—amid white limestone mountains—is still covered with crushed and thoroughly burned and blackened rocks.

Figure 109

The obliteration of the "sinning cities" was a muddled affair. According to the Sumerian texts, Ninurta tried to dissuade Nergal from carrying it out. According to the Bible, it was Abraham who pleaded with one of the three Angels who had dropped-in on him to spare the cities if as few as ten "righteous ones" shall be found in Sodom. That evening, in Sodom, two Angels sent to verify whether the cities should be spared were mobbed by a crowd seeking to sodomize them. "Upheavaling" was inevitable; but they agreed to delay it in order to give Lot (Abraham's nephew) and his family enough time to escape to the mountains. Then at dawn,

> Erra, emulating Ishum,
> the cities he finished off,
> to desolation he upheavaled them.

Sodom and Gomorrah and three other cities in "the disobedient land, he obliterated." The Bible, in virtually identical words, relates that "as the sun was risen over the Earth, from the skies were those cities upheavaled, with brimstones and fire that have come from Yahweh."

> And Abraham got up early in the morning,
> and went to where he had stood with the Lord,
> and gazed toward Sodom and Gomorrah,
> in the direction of the place of the Plain;
> and lo and behold—
> there was steaming smoke rising from the ground
> like the steaming smoke of a furnace.
>
> Genesis 19:27–28

That is how it was, the Bible states, "when *the **Elohim*** annihilated the Cities of the Plain." Five nuclear devices, dropped by 'The Annihilator' Nergal, did it.

And then the Law of Unintended Consequences proved itself true on a catastrophic scale; for an unexpected consequence of the nuclear holocaust was the death of Sumer itself: A poisonous nuclear cloud, driven eastward by unexpected winds, overwhelmed all life in Sumer (Fig. 110).

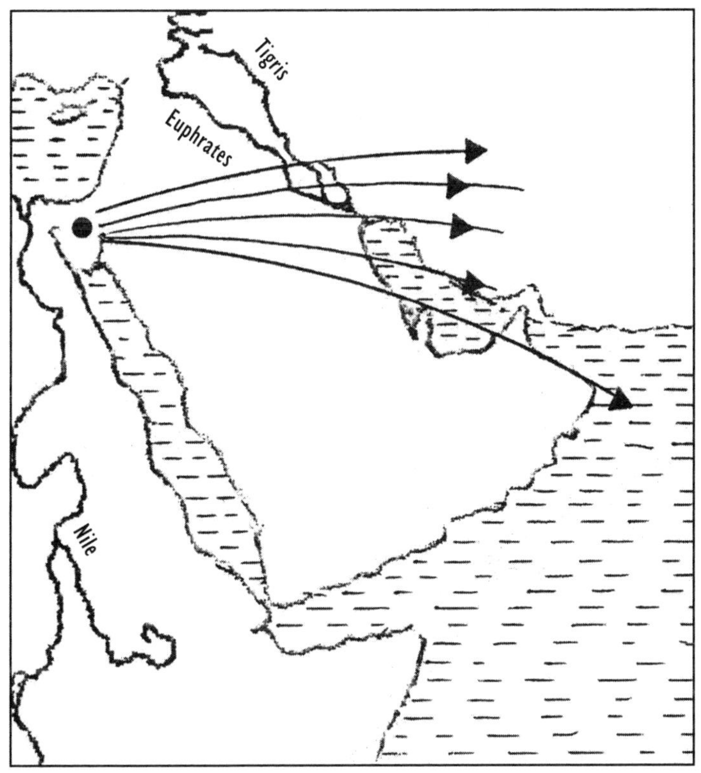

Figure 110

THE "EVIL WIND"

"A storm, the Evil Wind, went around in the skies, causing cities to become desolate, causing houses to become desolate, the sheepfolds to be emptied, causing Sumer's waters to be bitter, its cultivated fields grow weeds"—so did text after text from that time describe what had happened.

"On the land Sumer fell a calamity, one unknown to man, one that had never been seen before, one which could not be withstood," the texts say. An "unseen death roamed the streets, it let loose in the road . . . No one can see it when it enters the house . . . There is no defense against this evil which assails like a ghost; the highest wall, the thickest wall, it passes as a flood . . . through the door like a snake it glides, like a wind through the hinge it blows in . . . Those who hid behind doors were felled inside, those who ran to the rooftops died on the rooftops." It was a terrible, gruesome death: Wherever the Evil Wind reached, "the people, terrified, could not breathe . . . mouths were drenched in blood, heads wallowed in blood, the face was made pale by the Evil Wind."

It was not a natural calamity: "It was a great storm decreed from Anu, it had come from the heart of Enlil." ***It was the result of an explosion***: "An evil blast the forerunner of the baleful storm was." ***It was triggered by nuclear devices***—"caused by seven awesome weapons in a lightning flash"; and ***it came from the Plain of the Dead Sea***: "from the Plain Of No Pity it had come."

Forewarned of the Evil Wind's direction, the gods fled Sumer in panic. Long Lamentation Texts, such as *Lamentation Over The Destruction of Sumer and Ur,* list the cities and the temples that were "abandoned to the Wind" and describe the haste, panic, and grief as each deity fled, unable to help the people. ("From my temple like a bird I was made to flee," Inanna lamented.) Behind them temples, houses, animal stalls, all buildings remained standing; but everything alive—people, animals, vegetation—died. Texts written even centuries later recalled that day, when a cloud of radioactive dust reached Sumer, "The Day when the skies were crushed and the Earth was smitten, its face obliterated by the maelstrom."

"Ur has become a strange city, its temple has become a Temple of Tears," wrote a weeping Ningal in *A Lamentation Over the Destruction of Ur;* "Ur and its people were given over to the Wind."

XV
Buried in Grandeur

Four thousand years after the nuclear calamity, in A.D. 1922, a British archaeologist named Leonard Woolley came to Iraq to dig up ancient Mesopotamia. Attracted by the imposing remains of a ziggurat that stood out in the desert plain (Fig. 111), he chose to start excavating at the adjoining site locally called Tell el-Muqayyar. As ancient walls, artifacts, and inscribed clay tablets were unearthed, he realized that he was digging up ancient Ur—Ur of the Chaldees.

His efforts, lasting twelve years, were carried out as a Joint Expedition of the British Museum in London and the University of Pennsylvania Museum in Philadelphia. Some of those institutions' most dramatic exhibits consist of objects, artifacts, and sculptures found by Sir Leonard Woolley in Ur. But what he had found may well transcend anything that has ever been put on display.

As the arduous task of removing layers of soil that desert sands, the elements, and time accumulated over the ruins progressed, the contours of the ancient city started to emerge—here were the walls, there were the harbor and canals, the residential quarters, the palace, and the ***Tummal***—the artificially raised area of the sacred precinct. Digging at its edge, Woolley made the find of the century: A cemetery, thousands of years old, that included *unique 'royal' tombs.*

The excavations in the residential sections of the city established that Ur's inhabitants followed the Sumerian custom of burying their dead right under the floors of their dwellings, where families con-

Figure 111

tinued to live. It was thus highly unusual to find a cemetery, with as many as 1,800 graves in it. They were concentrated within the area of the sacred precinct and ranged in age from pre-dynastic times (before Kingship began) through Seleucid times. There were burials on top of burials, intrusions of graves into others, even instances of apparent re-interments in the same graves. In some instances, Woolley's workers dug huge trenches going down almost fifty feet, to cut through the layers and better date graves.

Most were hollows in the ground, where the bodies were placed lying on their backs. Woolley assumed that these different 'inhumations' were accorded on the basis of some social or religious status. But then, in the southeastern edge of the sacred precinct—within the walled area—Woolley discovered a group of entirely different burials, some 660 of them. In them, *with sixteen exceptions,* the bodies were wrapped in reed matting as a kind of a shroud, or placed in wooden coffins—an even greater distinction, for wood was in short supply and quite expensive in Sumer. Each one of those dead persons was then laid to rest at the bottom of a deep rectangular pit, large enough to hold them. The people thus buried, both male and female, were invariably placed on their sides—not on their backs as in the common burials; their arms and hands were flexed in front of their chests, their legs were slightly bent (Fig. 112). Laid out beside the bodies or on them were various personal belongings—jewelry, a cylinder seal, a cup or bowl; these objects enabled dating these graves to the Early Dynastic Period, roughly from circa 2650 B.C. to 2350 B.C.; it was the period in which central kingship was in Ur, starting with Ur's First Dynasty ("Ur I"), when Kingship was transferred thereto from Uruk.

Woolley reasonably concluded that the city's ruling elite was buried in these particular 660 tombs. But then Woolley unearthed the special sixteen tombs grouped together (Fig. 113) and landed an unprecedented find. They were entirely unique—unique not only in Sumer, but throughout Mesopotamia, throughout the whole ancient Near East; unique not only for their period, but for all periods. Clearly, Woolley

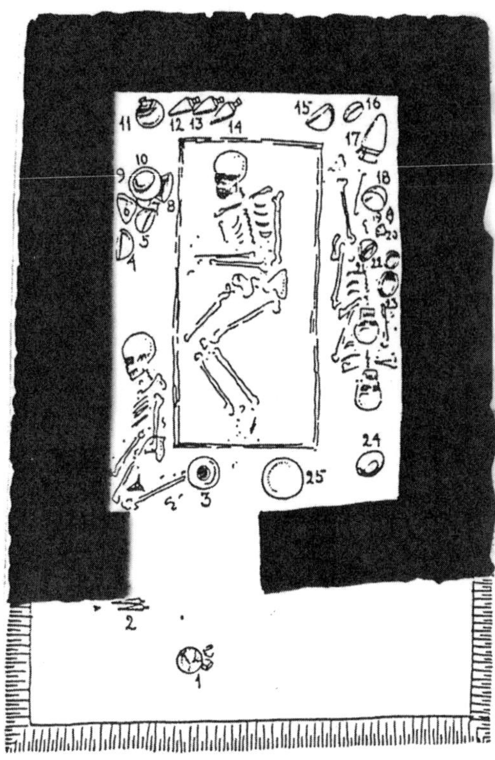

Figure 112

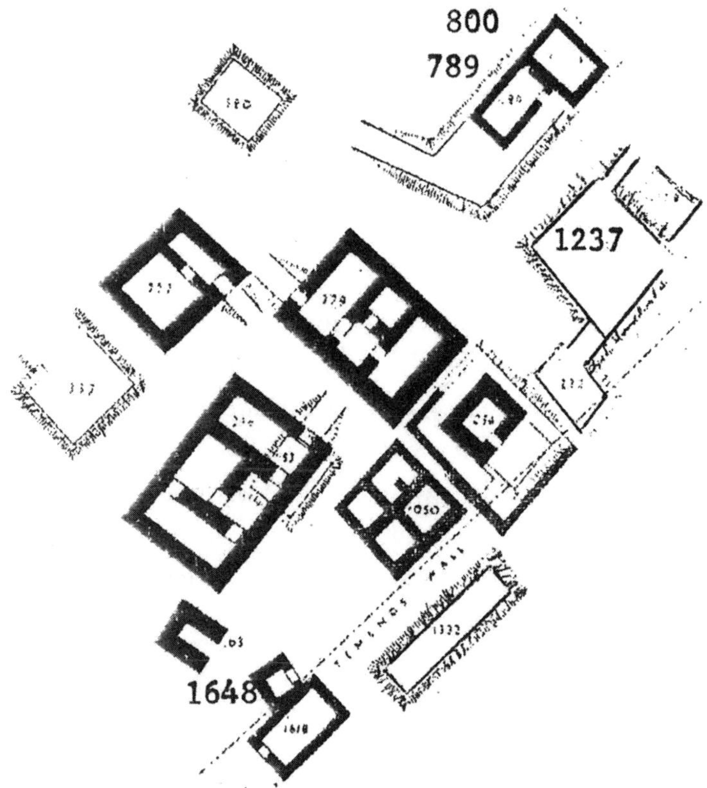

Figure 113

surmised, only someone of the highest importance had been buried in such special tombs and unique burials; and who was more important than the king or his consort, the queen? Cylinder seals in which names were combined with the titles ***Nin*** and ***Lugal*** convinced Woolley that he had discovered the ***Royal Tombs of Ur.***

His greatest single find was the tomb designated PG-800. Unearthing and entering it was an event in the annals of Mesopotamian archaeology comparable to the discovery and entering of Tut-Ankh-Amen's tomb in Egypt's Valley of the Kings by Howard Carter in 1922. To protect his unique find from modern robbers, Woolley notified his sponsors of the find by a telegram written in Latin; the date was January 4, 1928.

Subsequent scholars have accepted that logic and continue to refer to this unique group of tombs as the *Royal* tombs of Ur, even though some have wondered—because of what those tombs contained—who in fact was buried in several of them. Since to such scholars the ancient 'gods' were a myth, their bewilderment stops there. But if one accepts the reality of the gods, the goddesses, and the demigods—one is in for a thrilling adventure.

* * *

To begin with, the sixteen special tombs, far from being simple pits dug in the ground large enough to hold a body, were *chambers built of stones* for which a large excavation was made; they were set deep in the ground, and they had *vaulted or domed roofs* whose construction required engineering skills extraordinary for those times. To those unique structural features was added one more: Some tombs were accessible via well-defined sloping *ramps* that led to a large area, a kind of forecourt, behind which the actual tomb chamber was located.

Next to the exceptional architectural features, the tombs were unique in the fact that the body they held, lying on its side, was sometimes not just in a coffin but at times in a *separately constructed enclosure*. To all that was added the fact that the body was surrounded by *objects of extraordinary opulence and excellence*—in many instances, *one-of-a-kind* anywhere, any time.

Woolley designated the Ur tombs by a "PG" ('Personal Grave') code and number; and in a tomb designated ***PG-755***, for example (Fig. 114), there were more than a dozen objects beside the body in the coffin, and more than sixty artifacts elsewhere in the tomb. The objects included a splendid *golden helmet* (Fig. 115), a superb *golden dagger* in a magnificently decorated silver sheath (Fig. 116), a silver belt, a gold ring, bowls and other utensils made of gold or silver, gold jewelry with or without decoration with lapis lazuli (the blue gemstone prized in Sumer), and a "bewildering variety" (to quote Woolley) of other metal artifacts made of electrum (a gold-silver alloy), copper, or copper alloys.

All that was entirely amazing for its time, when Man's metallurgical acumen was just advancing from use of copper (that needed no smelt-

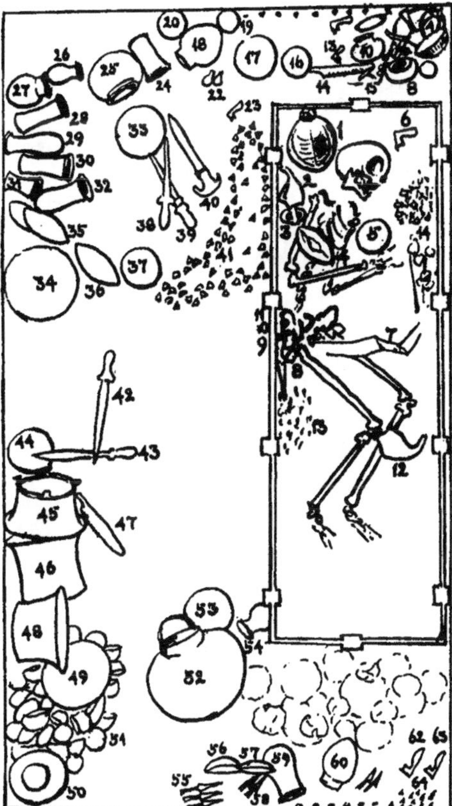

Figure 114

ing) to the copper-tin (or copper-arsenic) alloy we call Bronze. Objects of such artistry and metalworking techniques as the dagger and the helmet were absolutely unknown anywhere else. If these observations bring to mind the opulent golden death mask and magnificent artifacts and sculptures found in the tomb of Egypt's Pharaoh Tut-Ankh-Amen (Fig. 117), let it be remembered that he reigned circa 1350 B.C.—some twelve centuries later.

Other tombs contained both similar and different objects made of gold or electrum, all of outstanding craftsmanship. These included utensils of daily use, such as cups or tumblers—even a tube used for drinking beer—and *all were made of pure gold;* other cups, bowls, jugs,

Figure 115

Figure 116

Figure 117

and libation vessels were made of pure silver; here and there, some vessels were made of the rare alabaster stone. There were weapons—spearheads, daggers—and tools, including hoes and chisels, *also made of gold;* since gold, being a soft metal, deprived these implements of any practical use, those implements (usually made of bronze or other copper alloys) must have served only a ceremonial purpose, or were a status symbol.

There was a variety of board games (Fig. 118), and numerous musical instruments made of rare woods and decorated with astounding artistry, lavishly using gold and lapis lazuli for the decorations (Fig. 119);

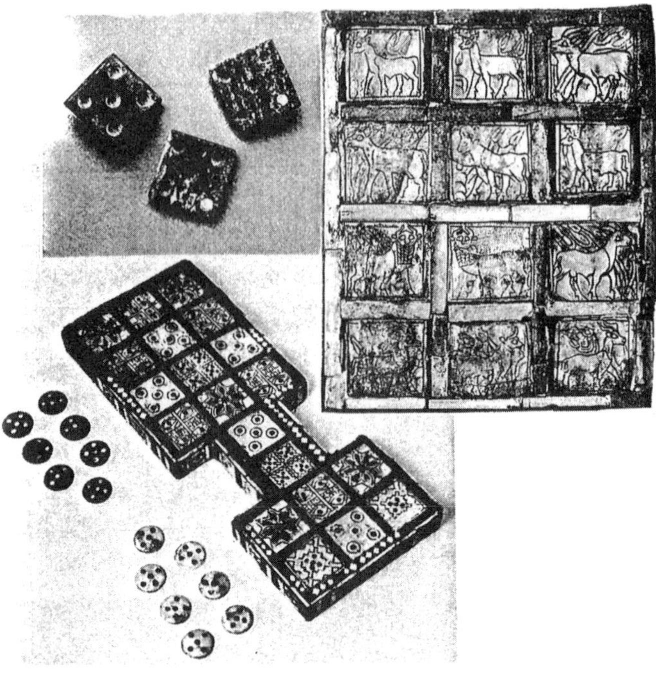

Figure 118

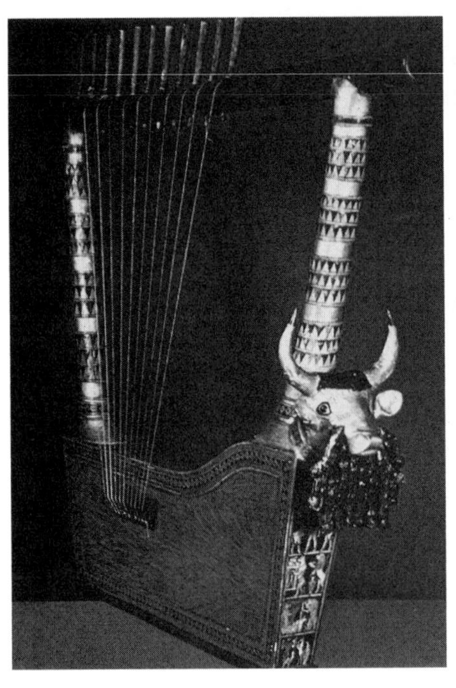

Figure 119

Figure 120

among them was a unique lyre made entirely of pure silver (Fig. 120). There were other finds, such as a complex sculpture (nicknamed 'The Ram in the Thicket', Fig. 121), that did not emulate any object or tool but were art for art's sake. For them, the artisans again lavishly used gold and combinations of gold with precious stones.

Similarly mind boggling was the array of jewelry, ranging from elaborate diadems and "headdresses" (a term employed by the archaeologists for lack of a better word) to chokers, bracelets, necklaces, rings, earrings, and other ornaments; they were all made of gold, semiprecious stones, or combinations thereof. In all of these objects, as in the ones mentioned earlier, the artistry and the techniques used to make and shape them—to create alloys, to combine materials, to weld them together—were

Figure 121

unique, ingenious, *and unmatched compared to any finds outside these tombs.*

One must bear in mind that none of the materials used in all of those objects—gold, silver, lapis lazuli, carnelian, rare stones, rare woods—were locally found in Sumer (or even in the whole of Mesopotamia). They were rare materials that had to be obtained and brought over from afar; yet they were used without any concern for rarity or scarcity. Above all there was the obviously abundant use of gold, even for the making of common objects (cups, pins) or tools (hoes, axes). Who had access to all those rare riches, who at a time when household utensils were made of clay or at best of stone, used uncommon metals for common goods? And who wanted everything possible to be made of gold, even if it rendered them impractical to use?

As one peruses records from those 'Early Dynastic' days, one finds that a king considered it a major achievement, by which that whole year was to be remembered, if he managed to have made and present to a deity a *silver* bowl—seeking in return prolonged life. Yet here, in selected tombs, myriad exquisitely crafted artifacts, utensils, and tools were made not just of silver but mostly of gold—an abundance and a use nowhere connected to royalty. Gold, it will be recalled, was the purpose of the Anunnaki's coming to Earth—to be sent back to Nibiru. In so far as an early and lavish use of gold *here on Earth and for common vessels* is concerned, we find gold mentioned only in inscriptions ***relating to Anu's and Antu's state visit to Earth circa 4000 B.C.***

In those texts, which were identified by their scribes as copies of original ones from Uruk, detailed instructions specified that *all the vessels used by Anu and Antu for eating, drinking, and washing "shall be made of gold";* even the trays on which food will be served had to be golden ones, as had to be the libation vessels and censers used in washing. A list of the variety of beers and wines that were to be served to Anu specified that the beverages had to be served in special ***Suppu*** ('Liquid holding') vessels made of gold; even the ***Tig.idu*** ('Mixing vessels') in which food was prepared had to be of gold. The vessels, according to those instructions, *were to be decorated with a 'rosette' design to mark them as 'Belonging to Anu'*. Milk, however, was to be served in special alabaster stone vessels, not in metal ones.

When it came to Antu, golden vessels were listed for her banqueting, mentioning the deities Inanna and Nannar (in that order) as her special guests; the *Suppu* vessels for them, as well as the trays on which they were served, also had to be of gold. All that, one ought remember, at a time *before* Mankind was granted civilization; so the only ones able to make all those objects had to be craftsmen of the gods themselves.

Remarkably, the Anu and Antu list of eating and drinking vessels that had to be made of gold, and in one specific case (for milk) of alabaster stone, *reads almost as an inventory of objects discovered in the 'royal' tombs of Ur;* so the questions 'Who had to have common utensils made of uncommon metals, who wanted everything possible to be made of gold?' led to "***The gods***" as an answer.

A conclusion that all those objects were for the use of gods, not mortal royalty, becomes more probable as we reread some of the Sumerian hymns to their gods—such as this one, inscribed on a clay tablet from Nippur (now languishing in the basement of the University Museum in Philadelphia). A hymn to Enlil, it extols his *golden hoe* with which he broke ground for the ***Dur.an.ki,*** the Mission Control Center in Nippur:

> Enlil raised his hoe,
> the hoe of gold with lapis lazuli tip—
> his hoe whose tied-on blade
> of silver-gold was made.

Similarly, according to the text known as *Enki and the World Order,* his sister Nin<u>h</u>arsag "has taken for herself the gold chisel and the silver hammer"—again utensils that, made of these soft metals, were only symbols of authority and status.

When it comes to the harp made of silver, we find that a rare musical instrument called ***Algar*** is specifically listed as one of Inanna's possessions in a 'Sacred Marriage' hymn to her by the king Iddi-Dagan: The musicians, it says, "play before thee the ***Algar*** instrument, of *pure silver* made." Though the precise nature of the instrument, which gave out "sweet music," is not certain, the ***Algar*** is mentioned in Sumerian texts as a musical instrument *played exclusively for the gods;* except that Inanna's was made of pure silver.

Such mentions of objects similar to those found in Ur's special tombs are found in other hymns; they become virtually countless when it comes to jewelry and such; and they are especially overwhelming when it comes to Inanna/Ishtar's jewelry and attire.

Yet as portentous as all that is, what was encountered in several of the 'Royal Tombs' was even more mind boggling; for even more unusual than the objects and opulence that accompanied some of the deceased was ***their accompaniment by scores of other human bodies buried along with them.***

* * *

Burials with others buried alongside the deceased were an unheard-of phenomenon anywhere in the ancient Near East; so the discovery of two 'companions' buried with the deceased in a tomb (designated ***PG-1648***) was already unusual. But what was found in some of the other tombs surpassed anything ever found before or thereafter.

Tomb ***PG-789,*** named by Woolley the ***'King's Tomb'*** (Fig. 122), began with a sloping ramp that led to what Woolley designated 'the Burial Pit' and to an adjoining burial chamber. Presumably, the tomb was entered and looted by grave robbers in antiquity, which may account for the absence of the main body and precious objects. But there were other bodies all over: Six 'companion' bodies were lying in the access ramp; they wore copper helmets and carried spears, as soldiers or bodyguards would. Down in the pit were remains of two wagons, each one drawn by three oxen whose skeletal remains were found *in situ* together with the bodies of one oxen-handler and two drivers per wagon.

All that was just a glimpse of what Woolley called "the king's retainers"—*fifty-four* of them, found in the 'Death Pit' (their precise locations marked by a skull sign in Fig. 122)—who, judging by the objects found near the bodies, were mostly males who held decorated spears with electrum spearheads; near them lay loose silver spearheads, reign rings made of silver, shields, and weapons; bulls and lions were a prominent feature of sculptures and decorations. While all that bespoke a military leader, the objects found near a smaller number of bodies identified as females bespoke appreciation of art and music: A sculpted bull's head made of gold with a lapis lazuli beard, wooden lyres exquisitely decorated, and a musical 'sound box' with panels

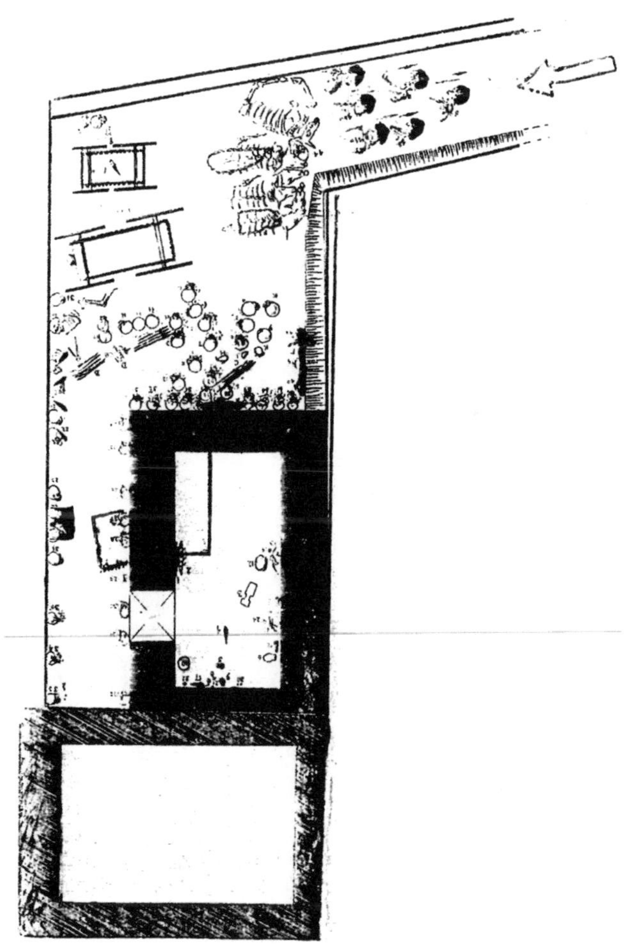

Figure 122

whose inlaid decorations depicted scenes from the tales of Gilgamesh and Enkidu.

An artist's 1928 rendition of what the assemblage in the death pit might have looked like, before everyone there was drugged or killed to be buried *in situ* (Fig. 123), gives a chilling reality to the scene.

Adjoining PG-789 was a similarly planned tomb, **PG-800**, that Woolley named "the ***Queen's Tomb.***" Here too he found accompanying bodies both in the ramp and in the pit (Fig. 124)—five bodies of guards, an oxcart with its grooms, and ten bodies presumably of female attendants who carried musical instruments. But *here there was a body lying on a bier, placed in a specially constructed burial chamber,* where it was accompanied by three attendants. This chamber was not robbed in antiquity, probably because it was a *secret sunken chamber:* Its roof, rather than its floor, was on the same level as the floor of the pit. Judging by the skeletal remains as well as by the profusion of jewelry, ornaments, and even a large wooden chest for clothing, it was the body of a female—the 'Queen', as Woolley called her.

Figure 123

Figure 124

The female's body was adorned—virtually from head to toe—with jewelry and accessories made of gold, gold-silver alloy (electrum), lapis lazuli, carnelian, and agate. Gold, and gold in combination with lapis lazuli and other precious stones, dominated these finds; gold and silver were the metals of which objects in daily use were made (with rare alabaster stone sometimes used for bowls); so were various artfully sculpted objects, such as the heads of a bull and of a lion. With a somewhat lesser opulence but similarly adorned were the female attendants who were buried with her: in addition to an elaborate golden headdress, each one was wearing golden earrings, chokers, necklaces, armbands, belts, finger rings, cuffs, bracelets, hair ornaments, wreaths, frontlets, and a variety of other adornments.

Near those two tombs Woolley found the forepart of another large tomb, ***PG-1237*** (see site map, Fig. 113). He unearthed the ramp and the pit, but did not find the burial chamber to which they must have belonged. He named the find "***the Great Death Pit***" because it contained *seventy-three* bodies of attendants (Fig. 125). Based on the skel-

Figure 125

etal remains and the objects found on or near the bodies, only five of them were males, lying alongside a wagon. Spread in the pit were *sixty-eight female bodies;* the objects found near them included an outstanding lyre (since known as the 'Lyre of Ur'), the 'Ram in the Thicket' sculpture, and a bewildering variety of jewelry. As in the other tombs, gold was the dominant material. (It was ascertained later that Woolley did find a burial chamber abutting PG-1237, but because the body in it was wrapped in reed matting, he considered it an intrusion from a later time and not the original burial.)

Woolley unearthed a few other 'death pits' without finding the burials with which they had been associated. Some, as ***PG-1618*** and ***PG-1648,*** held just a few bodies of what Woolley termed 'retainers'; others held many more: ***PG-1050,*** for example, held *forty bodies.* One must assume that they were all entombments essentially similar to PG-789, PG-800 (and probably also PG-755); and that intrigued scholars and researchers from Woolley on, for *these entombments had no parallel anywhere,* nor were they mentioned in the vast literary trove of Mesopotamia—with one exception.

A text dubbed *The Death of Gilgamesh* by its first renderer in English, Samuel N. Kramer, describes Gilgamesh on his deathbed. Informed by the god Utu that Enlil will not grant him eternal life, he is consoled by promises of "seeing the light" even in the Nether World where the dead go. Missing lines deprive us of the link to the final 42 lines, from which it could be surmised that Gilgamesh was going to retain in Nether World the company of "his beloved wife, his beloved son . . . his beloved concubine, his musicians, his entertainers, his beloved cupbearer," the chief valet, his caretakers, and the palace attendants who had served him.

A line (line 7 in the fragment's reverse side) that can be read to include the words "whoever lay with him in the Pure Place" or "When they had lain down with him in the Pure Place" is taken as an indication that *The Death of Gilgamesh* in fact describes an 'accompanied burial'—presumably an extraordinary privilege granted to Gilgamesh, who was "two-thirds of him divine," as compensation for not gaining the immortality of the gods. While this explanation of the legible lines remains debatable, *there is no escaping the uncanny similarity between the Death of Gilgamesh text and the stunning reality uncovered at Ur.*

Another recent debate whether the attendants, who were certainly part of the funeral procession, stayed to be buried voluntarily, were drugged, or perhaps killed as soon as they reached the pit, does not change the basic fact: There they were, demonstrating a most unusual practice, *unemulated and not practiced anywhere else* where kings and queens galore were buried over thousands of years. In Egypt, the 'Afterlife' notion included objects but not a host of co-buried attendants; the great Pharaohs were buried (amid an opulence of accompanying objects) in tombs hidden deep underground—lying by themselves in complete isolation. In the Far East, the buried Chinese emperor Qin Shihuang (circa 200 B.C.) was accompanied by an army of his subjects—but they were all made of clay. And though from A.D. times and on the other side of the world, we might as well mention a recent find in Sipan, Peru, of a royal tomb in which four bodies accompanied the deceased.

The Ur tombs with the death pits were, and remain, unique. So who was so special to be buried in such horrific grandeur?

Woolley's conclusion that the sixteen extraordinary tombs were of mortal kings and queens stemmed from the accepted notion that gods and goddesses were just a myth and did not physically exist. But the abundant use of gold, the extraordinary artistic and technologically advanced aspects of the objects, and other features that we have pointed out, lead us to conclude that ***demigods, and even gods, were buried there;*** and this finding is boosted by the discovery of *inscribed cylinder seals.*

* * *

Woolley's excavators found cylinder seals both inside tombs and away from them; several seals and seal impressions were found in a pile of discarded stuff that Woolley called the Seal Impression Strata, or SIS for short. All depicted some scene; some were inscribed with names or titles, identifying them as personal seals. If a name-bearing seal was found on or beside a body, it was logical to assume that it belonged to that person; and that could tell us a lot. The assumption has also been that the loose 'SIS' seals came from tombs that had been entered and looted in antiquity, the looters keeping valued objects and discarding 'valueless' pieces of stone; to modern researchers, even the SIS seals are invaluable; and we will use them as clues to be followed in ***unraveling the biggest mystery of the Royal Tombs: Who was buried in PG-800.***

On six of those seals the central depicted scene was of lions preying on other animals in the wild. One such seal was found in **PG-1382** (a one-person grave), another by the side of a sole skeleton in **PG-1054**. Though these seals left their owner's identities unknown, they did suggest that the owners were males with heroic attributes—an aspect that becomes evident from the third such seal, in which a wild man—or a man in the wilderness—was added to the depicted scene. It was found in **PG-261**, which Woolley described as a "simple inhumation that had been plundered." And this seal *had its owner's name inscribed on it* in clearly legible script (Fig. 126): ***Lugal An.zu Mushen.***

In his report Woolley did not dwell on this cylinder seal, though it plainly identified it as the tomb of a king. Subsequent scholars have

Figure 126

also ignored it because since ***Lugal*** meant 'king' and ***Mushen*** meant 'bird', the inscription makes little sense when read "King Anzu, Bird." The inscription, however, becomes highly significant if it is read—as I suggest—"King/Anzu Bird," for it will then suggest that the seal belonged to the *King of **'Anzu bird'** fame—**it would identify the owner as Lugalbanda,*** whose way to Aratta, the reader will recall, was blocked at a vital mountain pass by the monster ***Anzu mushen*** ('Anzu the Bird'). Challenged to identify himself, that is what Lugalbanda answered:

> *Mushen,* in the *Lalu* I was born;
> *Anzu,* in the 'Great Precinct' I was born.
> Like divine Shara am I,
> the beloved son of Inanna.

Could the demigod Lugalbanda—a son of Inanna, spouse of the goddess Ninsun, and the father of Gilgamesh—***be the VIP who was buried in the violated and plundered tomb PG-261?***

If we are right in suggesting so, other pieces of the jigsaw puzzle will begin to form a plausible picture never before contemplated.

Though no telltale golden objects were found in it, strewn about in PG-261 were (per Woolley) "remnants of an assemblage associated with military men"—copper weapons, a bronze ax, etc.—objects befitting Lugalbanda who came to fame as a military commander for Enmerkar. Since the tomb had been entered and plundered by ancient grave robbers, it could well be that there had been in it varied precious artifacts that were carried off.

To envision how PG-261 might have been originally, we can take a closer look at the very similar tomb PG-755, where the golden helmet and golden dagger were found (see Figs. 115, 116). We do know who owned them, because among the artifacts inside the coffin two gold bowls, one actually held by the hands of the buried occupant, were inscribed with the name ***Mes.kalam.dug***—the name, no doubt, of the buried person. His name, with the prefix ***Mes*** (= 'Hero'), as explained by us earlier, meant *'Demigod'*. Not 'deified' as Lugalbanda and Gilgamesh were, his name does not appear in the God Lists (in fact, the only instance throughout the God Lists of a name that begins with *Mes*—a partly legible name that reads Mes.gar.?.ra—is found among the sons of Lugalbanda and Ninsun). But Mes.kalam.dug (= 'Hero who the Land held') is not a complete unknown: We know that he was a king from a cylinder seal bearing the inscription ***Mes.kalam.dug Lugal*** ('Meskalamdug, *king*') that was found in the SIS soil.

We know something about his family: Metal vessels, lying near his coffin in PG-755, bore the names ***Mes.Anne.Pada*** and ***Nin.Banda Nin,*** suggesting that they were related to the deceased; and we know who Mes.anne.pada was: He is listed in the Sumerian King List as the all-important *founder of the First Dynasty of Ur*! And he did not earn this honor without the highest qualifications: As stated in a British Museum text that we have quoted earlier, his "divine seed giver" was Nannar/Sin himself. Being only a demigod meant that his mother was not Nannar's official spouse, the goddess Ningal; but his genealogy still made him a half-brother of Utu and Inanna.

We also know who, in this context, the female ***Nin.Banda.Nin*** was: A two-tiered cylinder seal (belonging to the 'man and animals in the wilderness' series) found in the SIS pile (Fig. 127) was inscribed ***Nin.banda Nin/Dam Mes.anne.pada***—*'Ninbanda, goddess, spouse [of] Mesannepada'*—identifying her as the spouse of the founder of the 'Ur I' dynasty.

How was Mes.kalam.dug related to this couple? While some researchers hold that he was their father (!), to us it is obvious that

Figure 127

a demigod could not have been the father of a *Nin*—a goddess. *Our guess is that Nin.banda-Nin, was the mother of Meskalamdug, and Mes.anne.pada was his father;* and we further suggest that the discovery of their seals in the SIS soil undoubtedly means that ***they too were buried in the 'Royal Tombs' group,*** in tombs that had been entered and robbed in antiquity.

It is at this point that one must clearly and emphatically put an end to the continued scholarly reference to Ninbanda as 'queen'. ***Nin,*** as in Ninḫarsag, Ninmaḫ, Ninti, Ninki. Ninlil, Ningal, Ninsun, and so on, was always a *divine* prefix; the Great God List includes 288 names or epithets whose prefix was ***Nin*** (sometimes also for male gods, as in Ninurta or Ningishzidda, where it indicated 'Lordly/divine Son'). ***Nin.banda was not a 'queen',*** even if her spouse was a king; ***she was a NIN, a goddess;*** as the inscription doubly stated, she was "***Nin****.banda,* ***Nin***"—confirming that Mes.anne.pada was her husband, and *leading to the conclusion that the VIP entombed in PG-755—Mes.kalam.dug—was the son of that goddess + demigod couple who started the First Dynasty of Ur.*

The relevant section in the Sumerian King List states that Mesannepada, the founder of 'Ur-I' dynasty, was succeeded on the throne in Ur by his sons ***A.anne.pada*** and ***Mes.kiag.nunna***. They both bore the ***Mes*** prefix, thereby confirming that they too were demigods—as of course they were if their mother was the goddess Nin.banda. The firstborn son, Mes.kalam.dug, is not included in the 'Ur I' list; his title *Lugal* suggests that he reigned elsewhere—in the family's ancestral city Kish.

Could it be that the only one of this group of 'Ur-I' kings who was 'royally' buried in Ur was Meskalamdug, the one who did not reign in Ur? Not only the discarded cylinder seals listed above, but also a damaged seal imprint (with the familiar heroic scene, Fig. 128) found in the SIS soil bearing the name ***Mes.anne.pada,*** founder of the dynasty, suggest *that ancient robbers found his grave,* robbed it, and threw away (or dropped) the seal that was with the body. Which grave? There are enough unidentified tombs to choose from.

Figure 128

As the jigsaw puzzle of the first 'Ur-I' family and its burials emerges, it behooves us to wonder who the mother—Nin.banda-Nin—was. Was there a connection between ***Lugal.banda*** ('Banda the king') and ***Nin.banda*** ('The Goddess Banda')? If Lugal.banda, as we have suggested, was buried in Ur, as were Nin.banda's spouse, Mes.anne.pada, and three sons—what happened to her? Did she, with her Anunnaki longevity, need no burial—or ***did she herself, at some point, die and was also buried in this cemetery***?

This is a questions to be kept in mind as we unfold, step by step, the amazing secret lurking in the Royal Tombs of Ur.

* * *

The sixth 'wilderness scene' cylinder seal that depicts a crown-wearing naked male bears a clear inscription of its owner: ***Lugal Shu.pa.da***

(Fig. 129), 'King Shupada'. We know nothing of him except that he was a king; but that fact alone is significant, because the seal was found next to his body *in the pit of PG-800, where he was one of the male attendants*. Depicting him naked would be in line with earlier instances in which a naked ***Lu.Gal*** served a female deity (see, for example, Fig. 77).

That a *king* served as a funeral retainer makes one wonder whether the other grooms and attendants and musicians, etc., who accompanied the deceased VIP were mere servants, or rather high officeholders and dignitaries in their own right. That the latter is the case is additionally suggested by another find, near the wardrobe chest in PG-800, of a seal bearing the identification ***A.bara.ge,*** which can be translated 'The Water Purifier of the Sanctuary'—the personal seal of an officeholder who, as the deity's cupbearer, was the deceased's most trusted personal aide.

That the attendants of entombed VIPs were high-ranking persons in their own right is further attested by a cylinder seal found in the Great Death Pit of PG-1237. Depicting females banqueting and having beer with drinking straws while musicians are playing (Fig. 130), it belonged to a female courtier and was inscribed ***Dumu Kisal***—'Daughter of the Sacred Forecourt'. This too was a title of no small import, for it linked the title of its holder to a subsequent king named ***Lugal.kisal.si*** (= 'The

Figure 129

Righteous King of the Sacred Forecourt'), indicating her royal-priestly genealogy.

While PG-755 yielded an entombed body without its death pit, PG-1237 a death pit without a grave and a body, and PG-789 (the 'King's Tomb') a grave and its pit but no body, PG-800 emerged as the ideal discovery, providing the archaeologists with a body, a grave, and a death pit. Understandably, in Woolley's and all other researchers' opinions, PG-800 was "the richest of all the burials" in the Royal Cemetery of Ur. He also viewed the 'King's' PG-789 and the 'Queen's' PG-800—which sat right against each other—as a special unit, similar in their having the sloping ramp, the bier or coffin carrying wagon, the death pit filled with attendants who themselves were high ranking, and the special separate Tomb Chamber constructed as an underground stone building.

Whoever was buried in such a 'with pit' tomb with attendants who were themselves VIP's—even a king—had therefore to be more important than a mere royal princess or a king; it had to be at least a demigod—or even ***a fully qualified god or goddess.*** And that leads us to *the*

Figure 130

greatest enigma of the Royal Tombs of Ur—the identity of the female who was laid to rest in PG-800.

* * *

We can start unraveling the mystery by taking a closer look at the objects and adornments found with her. We have already described some of the golden abundance in PG-800 (which was not robbed in antiquity), extending to the fashioning out of gold even of utensils in daily use—a bowl, a cup, a tumbler—and we noted the similarity of such use to the specifications for Anu and Antu's stay in Uruk some two thousand years earlier.

The similarity additionally embraces Anu's emblem, the 'rosette' of flower leaves; so it is not without great significance *that the same symbol has been found embossed into the bottom of the golden utensils in PG-800* (Fig. 131). This could be possible if the utensils found in Ur were the very same ones from Anu's visit at Uruk, somehow preserved for two millennia as a family heirloom—in this case a feat linked to Inanna, to whom Anu bequeathed the E.Anna temple in Uruk with all in it. If the utensils were made afresh in Ur, then the VIP for whom they were made had to be entitled to display Anu's symbol. Who could that be, other than someone *directly* belonging to Anu's dynastic family?

Another clue, in our opinion, is an inconspicuous object found in

Figure 131

PG-800—a pair of golden 'tweezers'. The archaeologists assumed that it was made for cosmetic use. Maybe. But we find an identical object depicted on a cylinder seal that (according to its inscription) belonged to a Sumerian ***A.zu,*** a physician. We show the 'tweezers' from PG-800 superimposed on the cylinder seal (Fig. 132) to support the conclusion that it was *a medical instrument.* We don't know whether this symbolic emulation in soft gold indicated the profession of the deceased or was also an inherited family heirloom; in either case, it ***suggests that the goddess in PG-800 had links to a medical tradition.***

We now come to the jewelry and adornments of the buried "Queen" (as Woolley called her). Every detail about them justifies the adjectives 'unusual', 'remarkable', 'extraordinary'; they definitely deserve extra attention.

She was laid to rest wearing on her torso not a dress, but a cape made entirely of beads (Fig. 133). As already mentioned, there was a large 'wardrobe chest' outside the tomb chamber, indicating that the

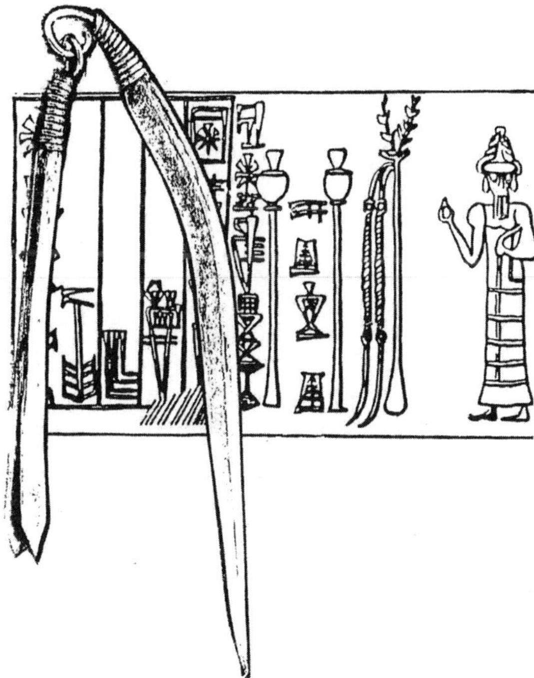

Figure 132

'queen' had ample clothing. Yet from the neck down the naked body was bedecked not with a garment but with long strings of beads—sixty of them—made of gold combined in artistic designs with lapis lazuli and carnelian beads. The strings of beads formed a 'cape' that was held in place at the waist by a belt made of golden strings decorated with the same gemstones. There were gold rings on each of her ten fingers, and a golden garter that matched the belt was worn on her right leg. Nearby, on a collapsed shelf, lay a diadem of gold and lapis lazuli adorned with rows of miniaturized animals, flowers, and fruits, all made of gold. Even the pins were artfully made of gold.

Undoubtedly, the most glittering and eye-catching of her accoutrements was the large and elaborate headdress the 'queen' wore. It was found crushed by fallen soil and was restored and placed by experts on a model's head (Fig. 134); it has since been among the best known and

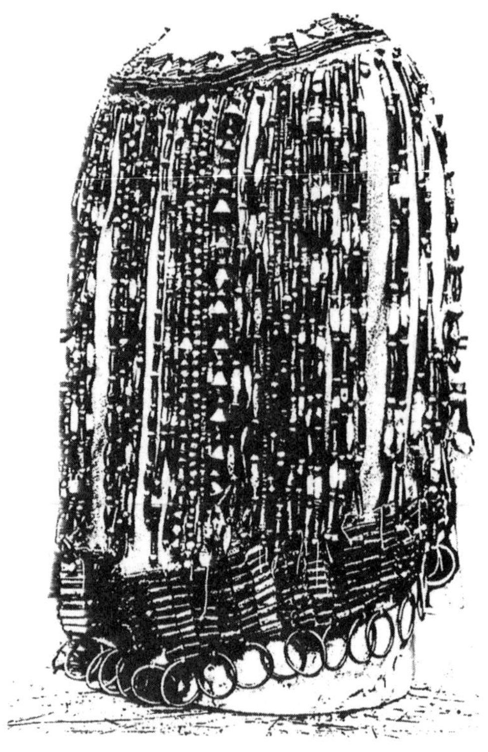

Figure 133

Figure 134

most exhibited objects from the Royal Tombs of Ur. Facing the entrance to the Sumerian Hall in the University Museum in Philadelphia, it usually evokes a 'Wow!' reaction on first sight. That too was my reaction the first time I saw it; but having become familiar with it and where it was found, it seemed odd that the only way to fit it on the head of a mannequin (made to resemble female heads found at Sumerian sites) was by artificially giving the mannequin an immense coif of stiff hair. The weighty headdress was held in place with golden pins and golden ribbons; matching its design and size were huge golden earrings adorned with precious stones.

The disproportion of the headdress is obvious when one looks at the golden headdresses worn by the female attendants who were buried with the 'queen' (Fig. 135). Similar to hers but less elaborate, they fitted perfectly on the heads without resort to a mass of artifical hair. *So either the 'queen' wore a headdress that was not hers—**or she had an unusually large head**.*

Figure 135

The 'queen' wore around her neck a choker, a collar, and a necklace, all made of gold combined with gemstones. The choker had at its center a golden rosette (the emblem of Anu); the collar bore a design that consisted of a series of alternating triangles, one of gold, the other of lapis lazuli (Fig. 136, top row); chokers or collars with the same design were also found worn by some of the female attendants in PG-1237 (bottom rows). This is highly significant, for in some of her depictions the goddess Inanna/Ishtar (superimposed image) was shown wearing *the exact same collar*! The exact same design was also deployed at the entranceway and on ceremonial columns (Fig. 137) in the earliest Ninmah/Ninharsag temples. Apparently reserved for female deities, this 'cult design' (as scholars call it) suggests some kind of affiliation between the several goddesses involved.

These and previous link-points to Inanna call for a closer look at both the unique bead cape and the exceptional headdress worn by the 'queen' in PG-800. The profuse use of lapis lazuli and carnelian requires reminding that the nearest source for lapis lazuli was Elam (nowadays

Figure 136

Figure 137

Iran), and carnelian was found farther east, in the Indus valley. As told in the *Enmerkar and the Lord of Aratta* text, it was to adorn Inanna's abode in Uruk that the Sumerian king demanded from Aratta tribute of carnelian and lapis lazuli. So it is not without significance that one of the few art objects found in the ruins of the Indus Valley centers, a statuette of Aratta's goddess—Inanna—depicts her naked and bedecked only with strands and necklaces of beads and golden pendants, held in place by a belt with a disc-emblem (Fig. 138). The striking similarities to the 'queen' in PG-800 with her beaded cape and belt do not end there: ***the statue's towering headdress with its large earrings looks as though an artist tried to emulate in clay the headdress in PG-800***.

Does it all mean that the 'queen' entombed in PG-800 was the goddess Inanna? It could have, were it not for the fact that Inanna/Ishtar was alive centuries later, when the Evil Wind overwhelmed Sumer; we know that because she and her hurried escape are clearly described in the Lamentation Texts. And she was also active many centuries later—into Babylonian and Assyrian times, in the 1st millennium B.C.

But if not Inanna—who?

Figure 138

WHEN 'IMMORTALS' DIED

The 'Immortality' of the Anunnaki gods, we have already observed, was in reality a great longevity that can be attributed to their Nibiruan life cycle. The notion of gods (or even demigods) as immortals has come to us from Greece; the discovery of Canaanite 'myths' at their capital Ugarit (on Syria's Mediterranean coast) showed where the Greeks got the idea.

By listing the ancestor couples on Nibiru, the Anunnaki acknowledged that they were long dead. In the very first 'Paradise' tale of Enki and Ninmah, she afflicts him with maladies (to stop his sexual shenanigans) that bring him to the brink of death—allowing that gods can get sick and die. Indeed, the very arrival of Ninmah the doctor and her group of nurses admits illness among the Anunnaki. The deposed Alalu, swallowing Anu's 'Manhood', died of poisoning. The evil Zu was captured and executed.

Sumerian texts described the death of the god Dumuzi, who drowned when escaping from Marduk's 'sheriffs'. His bride, Inanna, retrieved his body, but all she could do was mummify it for a hoped-for future resurrection; various later texts refer to Dumuzi as a resident of a 'Netherworld'. Inanna herself, going uninvited to her sister's Lower World domain, was put there to death—"a corpse, hung from a stake." Two android rescuers retrieved her body and with a 'Pulser' and an 'Emitter' brought her back to life.

When the nuclear Evil Wind began blowing toward Sumer, the gods and goddesses—neither immune nor immortal—hurriedly escaped in panic. The god Nannar/Sin tarried, and was afflicted with a limp. The great goddess Bau of Lagash refused to leave her people, and the Day of the Calamity was her last day: "*On that day, **as if she were a mortal,** the Storm by its hand seized her,*" a lamentation text states.

In the Babylonian version of *Enuma elish*—that was *read publicly* during the New Year festival—a god named ***Kingu*** (namesake of the leader of Tiamat's host) is killed to obtain blood for Man's creation.

In Sumer, the death of gods was as accepted as tales of their being born. The question is, Where were they buried?

XVI
The Goddess Who Never Left

Our question, 'Who was buried in PG-800?' would have sounded strange to Sir Leonard Woolley were he still alive to hear it. For as soon as he had reached its burial chamber—on January 4, 1928—he sent to the University Museum in Philadelphia a Western Union telegram that said (in translation from the Latin that he used for secrecy):

> *I found the intact tomb stone built and vaulted over with bricks of Queen Shubad adorned with a dress in which gems flower crowns and animal figures are woven together magnificently with jewels and golden cups. Woolley.*

"***The intact tomb of Queen Shubad.***" How did Woolley know this answer to the mystery as soon as he had found the chamber? Did the buried VIP have a name tag saying "Queen Shubad"? Well, in a manner of speaking, she did: Four cylinder seals were found in PG-800, one near the wardrobe chest and three inside the tomb chamber, depicting females banqueting. One of the three near the body was inscribed with four cuneiform signs (Fig. 139) that Woolley read ***Nin.Shu.ba.ad*** and translated '*Queen Shubad*'—for though ***Nin*** signified 'goddess', Woolley took it to mean 'queen', because as he and everyone else knew, gods and goddesses existed only mythically and had no physical body to be buried. His assumption that this was the personal seal of the buried VIP has been taken for granted, though the reading of her name has since been changed to ***Nin-Pu.a.bi***. (It is noteworthy that the University Museum in Philadelphia, on reopening the Royal Tombs of Ur exhibit in March 2004, changed the title from 'Queen Puabi' to '*Lady* Puabi'.)

The scene depicted on this seal, in two 'registers' is that of females banqueting; since tumblers are shown raised by the celebrants, they were probably drinking wine. In each register, there are two seated female celebrants and several female attendants/servants. The second and third seals found inside the tomb chamber also depicted, in two registers, two female celebrants—drinking beer with long straws, or having wine and food, served by attendants and entertained by a harp player. None of these two seals had any writing.

The fourth cylinder seal, found lying against the wardrobe chest outside the tomb chamber, also depicted banqueting scenes, with female celebrants and attendants. We have already pointed out that the name inscribed on it, ***A.bara.ge*** (= 'The Water Purifier of the Sanctuary') identified its owner as a high-ranking holder of the office of Cupbearer. We can additionally note here that he or she had to be a 'royal' per se, for he/she was a namesake of a famed king of Kish, ***En.me.bara.ge.si***—

Figure 139

a demigod who was credited with reigning 900 years (see chapter 11).

Apart from suggesting that the VIP buried in PG-800 was 'Queen Shubad', Woolley had no information to offer about her. There is no mention in Mesopotamian records of a queen by that name (whether ***Shubad*** or ***Puabi***). In so far as she was a **Nin**—a goddess—named ***Puabi,*** there is no such name in the God Lists either. If not an unlisted epithet—of which each deity had galore—it could have been a local or family nickname; so we will have to resort to detective tactics to unravel her identity.

The script-sign for ***Nin*** on the seal is absolutely clear and requires no further elaboration (see Fig. 57). Breaking down the epithet-name **Pu.a.bi** by its components, we find that the first one, read **PU**, was written with sign number 26a in the Sumerian Sign List—and it was another word for ***Sud***—'One who gives succor'—a nurse, a medic. This finding reinforces our earlier conclusion based on the medical 'tweezers' that the VIP entombed in PG-800 was a healer, as Ninmaẖ/Ninẖarsag, Ninlil (Enlil's spouse) and Bau (Ninurta's spouse) were; and ***our guess is that she was directly related to one of them,*** and thus an Enlilite.

The second component, read **Å** as in cuneiform sign number 383, meant 'Large/Much'; and **BI**, sign number 214, meant a certain variety of *beer.* So ***Nin Pu.a.bi*** literally meant a **Nin**, a goddess, who was "Healer [of] Much Beer." *It is a nickname that matched the banqueting and beer drinking* depicted on the the second cylinder seal found near Puabi's body (Fig. 140), Indeed, the depictions on all of the six 'female' seals found in the Royal Tombs show banqueting ladies who differ in certain aspects—age, hairdo, dress, and stature. Since the seal cutters might have tried to make the individual seals as true portraits as possible, these small details deserve attention. Especially intriguing is the PG-800 seal (see Fig. 139) in which, in the upper register, a younger goddess (the host?) sitting on the right next to the inscribed title/name, and a more matronly goddess, more elegantly dressed, and with an elaborate hairdo (the guest?) sits on the left. Was this an actual portrait of the tomb's occupant and her more matronly and hefty guest?

Figure 140

It's a possibility to be kept in mind, for the physical size of the hostess (and her guest) are relevant to their ultimate identification because some of the skeletal remains from several Ur tombs, including PG-800 and PG-755, were examined by the then leading British anthropologist, Sir Arthur Keith.

In regard to Shubad/Puabi, this is how he began his written report that formed part of Woolley's 1934 book on the Royal Tombs of Ur:

> An examination of the Queen's remains has led me to form the following conclusions concerning her:
> The Queen was about forty years of age at the time of her death;
> she was approximately 1.510 m. (5 feet) in stature;
> her bones were slender and her feet and hands small;
> she had a large and long head.

In estimating her age, Sir Arthur was baffled by the fact that dental and other aspects of her skeletal remains indicated a much younger age than forty. As to her stature, let us note that it is comparable to that of Inanna in the Mari photograph, Fig. 86.

While the skull, badly fractured, might have been compressed by soil pressure to appear longer and narrower than what it really had been, Sir Arthur concluded on the basis of detailed measurements that

the queen could not be a Sumerian—that she was "a member of a highly dolichocephalic race" ('dolichocephalic' is having a head disproportionately longer than it is wide). Even more so, he was astounded and puzzled by ***the overall size of the head and the extraordinarily large cranial (brain) capacity:***

> We have only to measure the frontal, parietal, and occipital bones along the midline of the vault to realize how large the capacity of the skull must have been . . .
> The cranial capacity could not have been less than 1600 cubic cm.—250 c.cm. above the mean for European women.

"The remains," he wrote, "***left no doubt*** that ***the Queen had an uncommonly capacious skull.***" After providing details of the rest of her bone remains, Sir Arthur's overall conclusion was that ***her head was unusually large,*** while her body, hands, and feet, compared to the size of the head, was rather small "though stoutly made."

To use Sumerian terminology, one can say that she had the head of a ***Gal*** and the body of a ***Banda*** . . .

Sir Arthur also examined the skeletal remains of the male in PG-755, referring to him as "Prince Mes-kalam-dug." Comparing the two, he observed that "except for her large cranial capacity, Queen Shub-ad was intensely feminine in her physical characterization; in Mes-kalam-dug the bones in the body were fashioned as a very robust male." His bones were much thicker than hers; "the right arm was particularly thick and strong in the Prince." All told, Sir Arthur concluded, "the bones of the Prince—alas! all of them are only fragmentary now—show him to have been a strongly built, powerful man, about 5 ft. 5 in. or 5 ft. 6 in. (1.650–1.675 meters) in height . . . He was a strong-necked man."

The skull of the 'Prince' had "the exact same cephalic index as in Queen Shub-ad" (i.e., the length to width proportion)—markedly elongated—and the cranial capacity (the brain size) was "well above the average size for Sumerians." Racially, Sir Arthur wrote, "I would name him, for lack of a better [word], Proto-Arab."

Fractured skull and bone remains from several other Early Dynastic tombs were examined; Sir Arthur's main conclusion was that they too were "Proto-Arabs." In an overall summary, he noted that the remains of the 'Queen' and the 'Prince' stood out from the others:

> Of particular interest is it to observe the fine physique and the rich brain endowments of Queen Shub-ad and of the Prince Mes-kalam-dug.
> The latter was an exceptionally strong man physically, and if we may rely on size of brain as an index of mental capacity—then was the Prince not only physically strong, but also a man of superior capacity.
>
> The Queen's cerebral endowment was exceptional, and if we can trust physical development of the body as a clue to sexual mentality, then we may infer that she was a very feminine woman.

In complete agreement with all the other aspects that we have found, Sir Arthur thus accurately described

- ***A heroic demigod in PG-755, a "strongly built powerful man" with a "superior cerebral capacity"***

and was right on the mark about the

- ***"Very feminine" smallish 'Queen' with an "uncommonly capacious skull" in PG-800.***

The skeletal remains and physical findings concerning the 'Prince' in PG-755 completely fit his identification as Mes.kalam.dug, whom we have established as a son of the goddess + demigod couple who began the First Dynasty of Ur; but we still face the enigma of the VIP in PG-800: Bejewelled and Inanna-like in stature, yet not Inanna . . . Who could she be, and who was entombed next to her in the emptied PG-789?

* * *

Regarding the occupant of PG-800 we have established the following points that can lead to her identification:

- A cylinder seal next to her body identified her as Nin.Puabi—***the goddess*** 'Puabi'.
- The retainers and attendants buried with her were themselves high-ranking courtiers, even a king, indicating that she was of greater importance than they—***that she was a goddess***—confirming her ***Nin*** title.
- Gold was used in this burial even for common, daily-use utensils—emulating the only other instance on record: Anu and Antu's visit to Earth ca. 4000 B.C.
- Those utensils were embossed with the same emblem—a 'rosette'—with which the Anu-visit utensils were embossed. This suggests that the female buried in PG-800 was 'Of the House of Anu'—***a direct linear descendant of Anu.*** Such a direct genealogical link to Anu could be ***through his sons* Enki *and* Enlil** or ***his daughters* Ninmah *and* Bau.**
- An implement found in the tomb, that has to be of the hardest metal—a hoe—was made of the soft metal gold, i.e., for symbolic purposes. The only recorded prior instance of that was the Sacred Hoe with which Enlil cut the ground to establish the *Duranki* Mission Control Center in Nippur. The hoe clue suggests that the ***VIP in this tomb was an Enlilite,*** associated with Nippur and not with Enki and Eridu. This eliminates Enki and ***leaves only three*—Enlil, Ninmah, *or* Bau—*as the direct genealogical link of 'Puabi' to Anu.***
- Possessing a symbolic golden medical instrument (the 'Tweezers') links Puabi to a tradition of giving medical succor—as **Ninmah and Bau** were; it still leaves the male Enlil in contention because his spouse, **Ninlil**, was also a nurse.
- Since it would seem improbable that the youthful-looking Puabi would have been one of the Olden Ones who had come to Earth from Nibiru, we ***cannot consider* Ninmah** or **Bau** or **Ninlil** themselves, and must look at ***their female descendants.***
- Since the known Earthborn daughters of Ninmah were fathered

by Enki, they are ruled out; we are left with ***daughters of*** **Enlil + Ninlil** ***or of*** **Bau + Ninurta**.

- Enlil + Ninlil had male sons (Nannar/Sin and Ishkur/Adad) born on Earth, and several daughters, including the goddess **Nisaba** (mother of king Lugalzagesi) and the goddess **Nina** (mother of king Gudea). Since Nina lived long enough to be one of the deities fleeing the later Evil Wind, ***she is eliminated*** as a 'Puabi' candidate. So does **Nisaba**, having still lived later, in Gudea's time.
- **Bau** (= 'Gula', the 'Big One'), youngest daughter of Anu, was married to Enlil's Foremost Son Ninurta. They had seven daughters of whom little is known except for **Ninsun**, spouse of the famed Lugalbanda; their famous son was Gilgamesh, so it had to be the mother, Ninsun, (rather than her smallish spouse) who bequeathed to Gilgamesh the physique of her father, Ninurta, and the heftiness of her mother, Bau/Gula.
- If claims by 'Ur-III' kings that Ninsun was their mother are valid, ***Ninsun herself could not be 'Puabi'*** (who was entombed during the 'Ur I' period).
- Going down the descendants' lines we arrive at the next Earthborn generation—a step in accord with Puabi being "in her forties" (per Sir Arthur Keith)—*if she were Earthborn.* The second Earthborn generation of known goddesses were Nannar/Sin's daughter **Inanna**, and a daughter of **Ninsun + Lugalbanda** named **Nin.e.gula**.
- Inanna (for reasons already given) could not be 'Puabi'. Yet Puabi's jewelry, beaded cape, the choker and its symbols, the all-silver harp, her great "femininity" (per Sir Arthur), etc.—and her stature—bespeak "*Inanna*"; so if Nin.Puabi was not Inanna herself, ***she had to be otherwise linked to Inanna.***
- Inanna had a known son (the god Shara) but no daughter; but she could—and did—have a granddaughter: Since Inanna, according to Lugalbanda's claim, was his mother, a daughter of Lugalbanda would have also been a ***granddaughter of Inanna,*** carrying her 'femininity' and love of jewelry traits.

- ***But the daughter of Lugalbanda would also be a granddaughter of Bau/Gula,*** for Lugalbanda's spouse, Ninsun, was a daughter of Bau + Ninurta!
- Her name (according to the God Lists) **Nin.e.gula** ("Lady of the House/Temple of Gula") serves as a confirmation that in addition to the 'femininity-jewelry gene' of Grandma Inanna she ***was bearing the 'Gula' gene*** of her grandmother Bau/Gula—the extraordinarily large head!

We thus obtain two genealogical-heritage lines of detection that converge:

Anu > ***Enlil*** + ***Ninlil*** > ***Nannar*** > ***Inanna*** > ***Lugalbanda*** + ***Ninsun***
and
Anu > ***Enlil*** + ***Ninmaẖ*** > ***Ninurta*** + ***Bau*** > ***Ninsun*** + ***Lugalbanda***

Thus converging, the two genealogical lines point to ***the same Lugalbanda + Ninsun couple as the progenitors of the goddess in PG-800: their daughter Nin.e.gula, also known as Nin.Puabi.***

This conclusion offers a plausible explanation for the contradictory physique of 'Puabi'—smallish body (a granddaughter of Inanna!) and an extraordinarily large head (a granddaughter of Bau/Gula).

This conclusion also offers a plausible reason for Lugalbanda to be the one entombed in PG-261.

And it explains the neglected clue of the naming of both ***Mes.Anne.Pada*** and ***Nin.Banda-Nin*** on vessels found near the coffin of Meskalamdug in PG-755, as well as in the seal inscription ***Nin.banda Nin/Dam Mes.anne.pada*** ('*Ninbanda, goddess, spouse [of] Mesannepada*'): Confirming, in our opinion, that they were the goddess + demigod couple who started the First Dynasty of Ur.

Does this solution of not only PG-800 but also of the other identifiable 'Royal' tombs make sense? Let's recall the intriguing fact that Ninsun has been involved in dynastic matchmaking—a glaring example having been her scheme to espouse one of her daughters to Enkidu. Was she beyond scheming, when the decision was made to transfer central

Kingship to a new dynasty in Ur, to have her daughter marry the demigod selected for the task? The other great matchmaking schemer, her mother, Bau/Gula—who might be the older matronly visitor shown on the cylinder seal having a cup of wine—would have given her blessing right away; and so would have the other grandmother, Inanna, for whom the choice represented a triumphant return to influence; was she the other visitor, sharing a beer?

Nin.banda, I suggest, was the daughter of Ninsun + Lugalbanda:

- ***Linked to Inanna by the dynastic title*** **Nin.banda**
- ***Given the epithet-name*** **Nin.e.gula** ***for her Bau heritage***
- ***Lovingly nicknamed*** **Nin.Puabi** ***for her constant partying***
- ***Laid to rest in the family's burial compound in the sacred precinct of Ur***

She was also, one realizes, a younger sister of Gilgamesh—both children of the unique couple: the deified demigod Lugalbanda and the mighty goddess Ninsun. And that opens up a wider subject.

* * *

While arriving at this (probable or at least possible) identification of the person in PG-800 is a gratifying achievement, an attempt to recognize identities in the other fifteen Royal Tombs is needed for understanding the jarring co-burials in the tomb chambers and especially in the death pits. The absence of any annals, hymns, lamentations, or other texts that would have explained the reasons is troubling in itself; the fact that the only textual corroboration is *The Death of Gilgamesh* text has only deepened the puzzle. But here is a thought outside-the-box: What if the Gilgamesh text described his *actual burial*—***what if the great Gilgamesh was actually buried in one of Ur's Royal Tombs?***

The burial place of **Gilgamesh** has never been found, nor do the available texts indicate where it was. All along it has been presumed that Gilgamesh was laid to rest where he had reigned—in Uruk; but nowhere in Uruk, a site that has been most extensively excavated, was such a tomb found. So why not consider the royal cemetery in Ur?

Transporting ourselves back to Sumer of almost 5,000 years ago, when central Kingship, having been in Kish and Uruk, was about to be transferred to Ur, we can imagine the chain of events that started in Kish. Beginning with the very first ruler, the kings were demigods: Mes.kiag.gasher was "a son of dUtu." So were the next ones—sons of a male god. To grasp the immensity of the change by the time of Lugalbanda, the father of Gilgamesh, it might be useful to reproduce a listing from an earlier chapter (to which one could add Gudea and his mother the goddess Nina):

Etana: Of same seed as Adapa (= Enki's)

Meskiaggasher: The god Utu is the father

Enmerkar: The god Utu is the father

Eannatum: Seed of Ninurta, Inanna put him on the lap of Ninharsag for breastfeeding

Entemena: Raised on Ninharsag's breastmilk

Mesalim: "Beloved son" of Ninharsag (by breastfeeding?)

Lugalbanda: Goddess Inanna is his mother

Gilgamesh: Goddess Ninsun is his mother

Lugalzagesi: Goddess Nisaba is his mother

Gudea: Goddess Nina is his mother

At first the kings are demigods by dint of being fathered by a male god and mothered by Earthling females (Enki himself having set the example in pre-Diluvial times). A transition, in which artificial insemination by a god but breastfeeding by a goddess, takes place. Then Lugalbanda enters the stage with a major change: From him on, the divinity comes from a female—*the mother is a goddess*. What we know now about DNA and genetics clarifies the significance of the change: The new demigods carry not only the mixed god-Earthling regular DNA, but also the second set of mitochondrial DNA that comes only from the mother. For the first time, in Lugalbanda, the demi-god is more than 'demi' . . .

What is to be done with **Lugalbanda** when he dies? He is more than a mere king, he is more than a usual demigod; but neither is he a real pure-blooded god, so he can't be taken to be laid to rest on Nibiru—nor can he be buried in Uruk's sacred precinct that has been sanctified by Anu himself. So the gods take him to Ur, the birthplace (and current residence) of his mother, Inanna. They 'deify' him by burying him at the edge of Nannar's sacred precinct in a specially built tomb—perhaps, as we have suggested, in **PG-261**—clutching his favorite ***Lugal An.zu Mushen*** seal.

Next, **Gilgamesh** appears on the scene, and he is also special: Not only is it his mother, not his father, who is the god-parent, but the father too is not a common Earthling: Lugalbanda, his father, was himself a son of a goddess (Inanna). So Gilgamesh is "two-thirds of him divine," enough to make him believe that he is entitled to the 'immortality' of the gods. Aided by his mother, the goddess Ninsun, and the god Utu in spite of their reservations, he goes on adventurous searches for eternal life that prove futile. Yet his conviction that he should not "peek over the wall as a mortal" continues even as he lies on his death bed—until Utu brings him the final verdict: Enlil said, No eternal life. But he is consoled: Because you are special, because you are unique, you shall continue to have with you your wife (and concubine . . .), cupbearer, attendants, musicians, and the rest of your household, even in the Nether-World.

And so—in this imagined scenario—Gilgamesh is buried near his father, in the sacred precinct of Ur, with the otherwise incomprehensible accompaniment promised him in lieu of Eternal Life. In which PG? We don't know, but there are several ones (emptied by ancient looters) to choose from. How about PG-1050 that held forty companion bodies—about the right number of those listed in *The Death of Gilgamesh* text?

An example is set—a precedent has been created.

With the death of Gilgamesh—we are now calendarwise circa 2600 B.C.—Uruk's heroic age peters out; all that remains of it are the epic texts and the depictions on cylinder seals, highlighting Gilgamesh, Enkidu, and heroic episodes. While the Anunnaki leadership contemplates where to site central Kingship, **Nin.banda**, *the sister of Gilgamesh,* and her spouse **Mes.anne.pada** mark time in Kish. As the decision comes that Ur was chosen, the goddess + demigod couple transfer there to assume the role of founders of the First Dynasty of Ur.

They leave behind in Kish their eldest son **Mes.kalam.dug**—reigning as King of Kish, though Kish is no longer the national capital. While the new rulers in Ur bring together Sumer's rivaling cities and extend Sumer geographically and culturally, their eldest son, Mes.kalam.dug, dies in Kish.

A demigod, he is laid to rest not far from his grandfather Lugalbanda and uncle Gilgamesh, in what is becoming the 'Ur I' dynastic family plot. Woolley, who designated the tomb **PG-755**, described it as a "simple inhumation" in which he found the deceased king's personal golden helmet and a magnificent golden dagger (found placed in the coffin beside the body). The more than sixty artifacts found in the tomb include personal objects (his silver belt, a gold ring, gold jewelry with or without lapis lazuli decoration) and his royal utensils, many of gold or silver—everlasting evidence of his demigod + royal status. But we really don't know whether a death pit was once part of a more elaborate burial—the fact that his personal seal inscribed ***Mes.kalam.dug Lugal*** ('Meskalamdug, *king*') was found discarded in the SIS soil does suggest that another, undiscovered part had existed and was entered and robbed in antiquity. Metal vessels, lying near the coffin in PG-755, bear the names of his parents ***Mes.Anne.Pada*** and ***Nin.Banda Nin***, further confirming the deceased's identity.

The day then comes when **Mes.anne.pada** himself "peers over the wall." His wife and two remaining sons provide him with an elaborate burial befitting the dynastic founder: A proper coffin, a stone-built tomb chamber, a death pit reachable via a sloping ramp. A great treasure of objects made of gold, silver, and gemstones was carried down along with the body on two wagons, each one drawn by three oxen and driven by two men and an oxen handler. Six soldiers wearing copper helmets and carrying spears acted as bodyguards. Down in the pit, many more soldiers were arrayed, carrying decorated spears with electrum spearheads and holding shields. A contingent of female singers and musicians was gathered with exquisitely decorated wooden lyres and a musical 'sound box' with panels whose inlaid decorations depicted scenes of the tales of Gilgamesh. Also brought down were varied sculptures decorated with images of bulls and lions; one particular sculpture, a favorite of the king, was that of a bull's head made of gold with a lapis lazuli beard. In all, fifty-four retainers assembled in the pit to keep Mes.anne.pada company in the Nether World.

When Woolley discovered this grave, he numbered it **PG-789** and called it the '***King's Tomb***'. He did so because of its obvious link to the 'Queen's' PG-800; and that, I suggest, in fact it was: ***The grave of Mes.anne.pada, the founder of the 'Ur I' dynasty.***

Because the main body was missing, and due to the absence of gold, silver, and lapis lazuli objects, Woolley concluded that PG-789 was entered and robbed in antiquity—quite possibly, when the digging for PG-800 revealed the tomb chamber of PG-789.

And so we arrive in our imagined Journey to the Past at 'Queen Puabi's' own death. How and when she died we do not know. Assuming she also outlived her two other sons (A.anne.pada and Mes.kiag.nunna) who reigned after her spouse had died, Nin.banda/Nin.e.gula/Nin .Puabi found herself alone, with all who were dear to her—her father Lugalbanda, her brother Gilgamesh, her spouse Mes.anne.pada, her three sons—dead and buried in the cemetery plot that she could daily see. Was it her wish to be buried on Earth alongside them—or could the Anunnaki not take her body back to Nibiru because, though a ***Nin,*** she did have some Earthly genes through her demigod father?

We don't know the answer. But whatever the reason, Nin.Puabi was buried in Ur, in a grave adjoining that of her spouse, with all the treasures and attendants to which this dynasty had uniquely become accustomed—adorned with jewelry from Grandma Inanna and an oversize headdress from Grandma Bau/Gula . . .

And that brings us to a Human Origins Discovery of all time: Because of all the Anunnaki and Igigi who had treaded planet Earth and were gone, ***Nin.Puabi—a NIN no matter who precisely she was—was The Goddess Who Never Left.***

NIN.PUABI'S DNA AND MTDNA LINEAGES

Here is how, if we are right, Nin.Puabi's general and specifically female DNA lines connect her directly to Nibiru, via Anu's children Enlil, Ninmah/Ninharsag, and Bau/Gula:

The Family Tree of Nin.Puabi

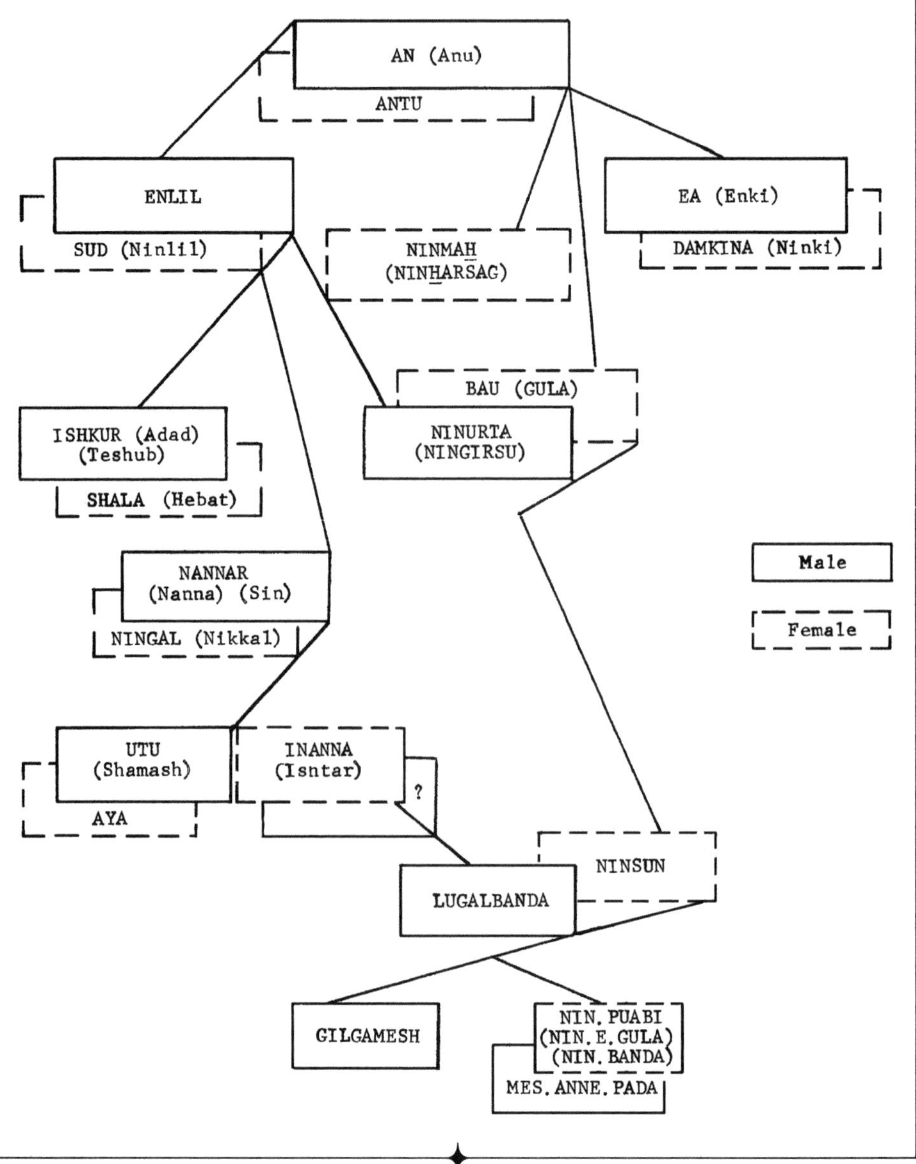

POSTSCRIPT

Mankind's Alien Origins: The Evidence

Ever since Darwin's offer of Evolution as the explanation for life on Earth, the most interesting chapter—that dealing with human origins—has crashed against two blocking walls as seawaves futilely striking a rocky shoreline: For the 'Believers', the sanctity of the biblical assertion that God, not Evolution, created Man; for the Scientific Purists, the inability to explain how, in a slow evolutionary process that requires millions and tens of millions of years, Man jumped from a hominid just learned to walk to Thinking Man (*Homo sapiens*)—us—practically overnight, some 300,000 years ago. The more ever-earlier hominid fossils are found, the greater the puzzle of the 'Missing Link' (as the problem has come to be known).

For more than thirty years now, since the publication of *The 12th Planet,* I have done my best to show that *there is no conflict* between Bible and Science, Faith and Knowledge. The 'Link' is missing, I said, because someone jumped the gun on Evolution and used sophisticated genetic engineering to upgrade a *Homo erectus* or *Homo ergaster* (as some prefer to call his African cousin) by mixing his genes with *their* advanced genes. That 'someone' were the biblical *Elohim* (whom the Sumerian called *Anunnaki*) who came to Earth from their planet, Nibiru, fashioned The Adam, then took the Daughters of Man as wives. That was possible, I explained, because life on their planet and on our planet is based on the same DNA—shared when the planets had collided . . .

Are you still there with me?

There has to be a better way, isn't there, not only to explain all that without arguments—not just to say that the Crime Scene Investigation indicates a murder had taken place—but a way to *produce the body and say:* ***Voila!***

Ah, if only one of the Anunnaki were still around, a chap or lass whose being Nibiruan would be unquestionable, who would roll up their sleeve and say: Test my DNA, decipher my genome, see that I am not of your planet! Find out the difference, discover the secret of longevity, cure your cancers . . . If only!

But, through the grace of Fate and the professionalism of dedicated archaeologists, such evidence—a physical body of an Anunnaki—does exist. It is the skeletal remains of Nin.Puabi.

It was in August 2002 that the British Museum in London revealed that unopened boxes languishing in its basement since Woolley's time contained skulls from the Royal Tombs of Ur. Seeking more information from the museum, I asked "whether there are plans to examine the DNA in these skulls." A polite reply informed me that "at present there are no plans to attempt DNA analysis," however "further research is conducted by the Department of Scientific Research and the Department of the Ancient Near East, and it is hoped that the initial findings may be made public early in 2003."

After further exchanges regarding the size of skulls and headdresses, the Curator of the Museum's Department of the Ancient Near East informed me that "a detailed reassessment of all the human bone collected from Ur is currently underway." The report, published in 2004, disclosed that the reassessment involved radiography (i.e., x-ray) tests by scientists at the Natural History Museum in London. It stated that "in spite of the long time since the skeletal remains were found, the conclusions of the contemporary specialists can be confirmed." The "contemporary specialists" in this case were Sir Arthur Keith and his aides.

Obtaining a copy of the report, I was astounded to realize that seventy years after Woolley' s discoveries, a museum in London still *possessed the intact skeletal remains of 'Queen Puabi' and 'Prince Meskalaindug'!*

Is this really so? I asked. Indeed so, the British Museum informed me on January 10, 2005: *"The skeleton of Puabi is held in the Natural History Museum, alongside others from Leonard Woolley's excavations at Ur."*

This was a bombshell discovery: The skeletal remains of a Nibiruan goddess (and of a demigod king) who was buried some 4,500 years ago—were unexpectedly available, intact!

One can debate who really built the Great Pyramids, disagree about the meaning of a Sumerian text, or dismiss an embarassing find as a forgery; but here is irrefutable physical evidence whose provenance, date and place of discovery, etc., are beyond doubt. ***So, if my identification of Puabi as an Anunnaki goddess and not a 'queen', and of Mes.kalam.dug as a demigod and not a Sumerian 'prince'—we have at our disposal two genomes of people fully or partly from another planet!***

Persisting with my repeated questions about whether DNA tests were or will be conducted, I was referred to the lead reassessment scientist, Dr. Theya Mollenson. By the time I could reach her, she retired. Attempts to find out more with the help of friends in London led nowhere. The need to deal with more pressing matters kept the issue on a 'back burner'—until recent news, that biologists were able to decipher and compare Neanderthal DNA from 38,000 years ago with that of modern man, struck as lightning: If so—why not decipher and compare the DNA of an Anunnaki female who died a mere 4,500 years ago?

In February 2009 I wrote about it to the Natural History Museum in London. A polite response signed by Dr. Margaret Clegg, Head of the Museum's Human Remains Unit, confirmed that their holdings include both "Nin Puabi, *also listed as Queen Shubad, and King Mes-Kalam-dug.*" Adding that "No DNA analysis has ever been conducted on these remains," she explained that "the Museum does not routinely conduct DNA analysis on remains in the collection, and there are *no plans to do so in the near future.*" This stance was reiterated by the museum in March 2010.

Though Nin.Puabi's DNA is not purely Anunnaki because her father, Lugalbanda, was only a demigod, her mitochondrial DNA that comes only from the mother is pure Anunnaki—leading through Ninsun and Bau to the Olden Mothers on Nibiru. If tested, her bones could reveal the DNA and mtDNA differences that represent our genetic Missing Link—that small but crucial group of "alien genes" (223 of them?) that upgraded us from wild hominids to Modern Man some 300,000 years ago.

It is my fervent hope that by showing that the remains of NIN-Puabi are no "routine" matter, this book will convince the museum to do the unusual and conduct the tests. They could provide vital explanation of the answer given to Gilgamesh:

When the gods created Man
Wide understanding they perfected for him;
Wisdom they had given him;
To him they had given Knowledge—
Everlasting life they had not given him.

What was it, genetically, that the 'gods' deliberately held back from us?

Maybe the Creator of All wished the Goddess Who Never Left to stay so that we finally find the answer.

ZECHARIA SITCHIN